DRAGON BONE HILL

DRAGON BONE HILL

An Ice-Age Saga of *Homo erectus*

NOEL T. BOAZ
RUSSELL L. CIOCHON

UNIVERSITY PRESS
2004

OXFORD
UNIVERSITY PRESS

Oxford   New York
Auckland   Bangkok   Buenos Aires   Cape Town   Chennai
Dar es Salaam   Delhi   Hong Kong   Istanbul   Karachi   Kolkata
Kuala Lumpur   Madrid   Melbourne   Mexico City   Mumbai   Nairobi
São Paulo   Shanghai   Taipei   Tokyo   Toronto

Copyright © 2004 by Oxford University Press

Published by Oxford University Press, Inc.
198 Madison Avenue, New York, New York 10016

www.oup.com

Oxford is a registered trademark of Oxford University Press

Library of Congress Cataloging-in-Publication Data
Boaz, Noel Thomas.
Dragon Bone Hill : an Ice-Age saga of *Homo erectus* / Noel T. Boaz,
Russell L. Ciochon.
p. cm.
Includes bibliographical references and index.
ISBN 0-19-515291-3
1. Peking man.
2. Excavations (Archaeology)—China—Zhoukoudian.
3. Zhoukoudian (China)—Antiquities.
I. Ciochon, Russell L.
II. Title.
GN284.7.B63 2004
569.9—dc21
2003012339

1 3 5 7 9 8 6 4 2
Printed in the United States of America
on acid-free paper

To Lanpo Jia (1908–2001)

Chinese paleoanthropologist and discoverer of Skulls X, XI, and XII.
His dedication to Dragon Bone Hill saved the site's excavation
data during World War II, and his regret was that he could not
do the same for the lost Peking Man fossils.
His ashes are buried at Zhoukoudian.

Contents

Preface

The coauthors of this book met in 1973 while they were both graduate students in paleoanthropology at F. Clark Howell's laboratory at the University of California at Berkeley. Although much of the lab's focus was then on Africa and Howell's Omo Research Expedition to Ethiopia, China was beginning to open up to renewed international paleoanthropological research. Howell was a member of the paleoanthropology delegation from the U.S. National Academy of Sciences to the People's Republic of China in 1975 and came back with news of great research possibilities. Ciochon was soon after to begin his own research projects in Asia, beginning with Burma in 1977, and extending over the next 25 years to India, China, Vietnam, Cambodia, and Indonesia. Boaz, on the other hand, continued his paleoanthropological research in Africa, working in Ethiopia, Libya, and the Democratic Republic of Congo. About ten years ago, however, their interests began to converge on the site of Zhoukoudian, also known as "Dragon Bone Hill" ("Longgushan" in Chinese). In 1993 Boaz met Professor Xiangqing Shao, a visiting physical anthropologist from Fudan University in Shanghai, China, in a graduate seminar he was teaching at George Washington University. Shao interested Boaz in renewed field research at Zhoukoudian, and after they had exchanged several letters with the Institute of Vertebrate Paleontology and Paleoanthropology in Beijing (IVPP), a joint research project began to take form. The ensuing agreement enabled the international and multi-institutional research on the Dragon Bone Hill site that Boaz and Ciochon have undertaken with Chinese colleagues, and which forms the basis of this volume. Professor Shao later also assisted Professor Alison Brooks of George Washington University in setting up an archaeological field school at Zhoukoudian before his untimely death in Washington, D.C., in 1999. Professor Shao is thanked for his role in furthering Chinese-American scientific cooperation and international paleo-anthropological research.

Our colleagues at IVPP in Beijing, Professor Qinqi Xu, former director of the Zhoukoudian International Research Center, and Jinyi Liu, were our coauthors on several professional papers published on this research. They were instrumental in planning our joint research, in constructing the Zhoukoudian excavation map, and in developing our collaborative taphonomic research of the extensive Zhoukoudian collections housed in Beijing and the Zhoukoudian Museum. Our January 1999 sojourn with them at Zhoukoudian was memorable for demonstrating to us what a chilly life it must have been for Peking Man in the Ice Age of northern China, and for how grateful we were for the amenities of the warm and hospitable Zhoukoudian guest house in which we stayed. Our many friends and colleagues at IVPP—Professors Xinzhi Wu, Wei Dong, Yamei Hou, Weiwen Huang, Wanbo Huang, and Yumin Gu, among others—are thanked for their many kindnesses and for their hospitality during our trips to China.

We owe a debt of gratitude to Professor Steve Weiner, chair of the Department of Environmental Sciences at the Weizmann Institute of Science, Rehovat, Israel, where Boaz spent the 1993–94 academic year as a Meyerhoff Visiting Professor. Applying Weiner's research methods—so successful in elucidating the geochemistry of traces of fire at Hayonim Cave in Israel—to the problem of fire at Zhoukoudian, seemed perfect. It was through Dr. Weiner's initiative that Dr. Xu went to Israel to learn the technique of X-ray analysis of sediments, which set the stage for the team of Weiner, Paul Goldberg, and Ofer Bar-Yosef to travel to China for the fieldwork that has so enlightened and informed our understanding of fire and the sedimentological history of Longgushan.

For access to collections and for productive and enjoyable discussions related to Asian *Homo erectus*, we thank Ian Tattersall, Eric Delson, Ken Mowbray, and Gary Sawyer of the American Museum of Natural History and our Indonesian colleagues Y. Zaim and F. Aziz. Over the years our discussions with G. H. R. von Koenigswald, F. Clark Howell, Sherwood Washburn, Phillip Tobias, Alan Walker, Geoff Pope, John Olsen, Milford Wolpoff, Philip Rightmire, Chris Stringer, John Fleagle, Alison Brooks, Rick Potts, Jack Cronin, Alan Almquist, Yoel Rak, and Robert Franciscus have contributed to the ideas presented in this volume. Peter Brown's paper at the 1991 "*Pithecanthropus*" symposium at the Senckenberg Museum in Frankfurt had a seminal effect on our thinking regarding *Homo erectus* cranial thickness. Chris Davett of the Washington State University Electron Microscope Center assisted with Scanning Electron Microscope analysis. Sandy Martin and Lynette Nearn are thanked for their significant contributions to our cranial pachyostosis studies. Christopher Janus, Lucian Pye, and Martin Taschdjian provided valuable insights into historical aspects of the disappearance and search for the Peking Man fossils.

We thank the following for their help in archival and library research for the project: Paula Willey of the American Museum of Natural History Library, New York City; and Ken Rose, Mindy Gordon, Darwin Stapleton, and Tom Rosenbaum of the Rockefeller Foundation Archives in Sleepy Hollow, New York. We owe special thanks to the staffs of the libraries at the University of Iowa (especially the interlibrary loan office), the Ross University School of Medicine, Old Dominion University, the Weizmann Institute of Science, Washington State University, Portland State University, the Portland (Oregon) Public Library, the University of California at Berkeley, Georgetown University (Walter Granger and Lucille Swan Collections), and the Smithsonian Institution (Frank Webb Collection).

John Olsen, Milford Wolpoff, and Robert Franciscus critically read the manuscript and we thank them for many valuable comments and suggestions. Rubén Uribe, Nathan Totten, Michael Zimmerman, and Erin Schembari helped with computer graphics. Wei Dong graciously scanned early photos of Zhoukoudian from the collections at the IVPP. Aidi Yin, M.D. and Yaoming Gu, M.D., assisted us in translating from Chinese. Jessica White commented on editorial issues. K. Lindsay Eaves-Johnson helped with editing the text and checking the bibliography. We thank our editors at Oxford University Press, Kirk Jensen and Clifford Mills, for their patience and valued assistance. Others who have assisted in forming our concepts and putting them into written form include Bruce Nichols, Le Anh Tu Packard, and Vittorio Macstro. We also acknowledge Le Anh Tu Packard for helpful comments on the final draft of the manuscript. We also thank agent Susan Rabiner for her help in promoting the project and Bill McCampbell for facilitating it. Finally, Meleisa McDonell, Lydia Boaz, Peter Boaz, Alexander Boaz, and Noriko Ikeda Ciochon are thanked for their patience and forbearance while this book was being written. Funding for Boaz was provided by the International Foundation for Human Evolutionary Research and the Ross University School of Medicine. Funding for Ciochon was from the dean of the College of Liberal Arts and Sciences, University of Iowa, and the Human Evolution Research Fund of the University of Iowa Foundation.

Authors' royalties from the sale of this volume will be donated to the Zhoukoudian Museum at Dragon Bone Hill, a United Nations World Heritage site.

Noel T. Boaz
Russell L. Ciochon

List of Illustrations

Color Illustrations Follow Page 76

DRAGON BONE HILL

CHAPTER 1

The Bones of Dragon Hill

In the 1920s, when the excavations started at Dragon Bone Hill, the understanding of human evolution was in a confused state. Eugene Dubois, the discoverer of *Pithecanthropus* from Java, was generally thought to have gone a bit insane in his advanced years. He had buried the fossils under his kitchen floor and had begun to think that he had discovered not the precursor of the human species but a giant gibbon-like primate instead. Henry Fairfield Osborn of the American Museum of Natural History was mounting a major expedition to Asia to look for the ancestors of humanity so far back in time that he ended up only with fossils of dinosaurs. A fossil tooth of an extinct pig-like peccary from Nebraska was, for a brief time, mistaken for an early humanlike ape in America and named *Hesperopithecus*. A fossil skull discovered by Raymond Dart in South Africa was named *Australopithecus* and claimed as a new human ancestor from that continent. And Professor Frederick Wood Jones of England was developing his elaborate albeit totally fallacious theory that humans had evolved directly from tarsiers—small, nocturnal, leaping primates now found only in Southeast Asia. Adding to this already rich tapestry of confusion was the "Piltdown Man" hoax—a modern human skull, a broken orangutan jawbone, and isolated teeth—planted in southern England and claimed by some to be humanity's oldest known ancestor. Out of this paleoanthropological morass there arose in the 1930s a clear ancestor—adroitly discovered, expertly studied, meticulously published, and universally acclaimed. It became widely known as "Peking Man." This book is about that hominid,[1] now known scientifically as *Homo erectus*.

For much of the first half of the twentieth century, the smart money was on Asia as the place of origin of the human lineage. Africa, a continent that

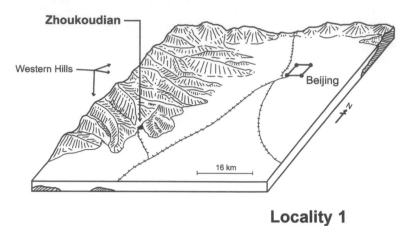

Zhoukoudian

Western Hills

Beijing

16 km

Locality 1

Zhoukoudian

SW

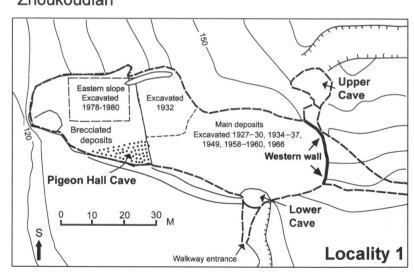

150

Eastern slope
Excavated
1978–1980

Excavated
1932

Main deposits
Excavated 1927–30, 1934–37,
1949, 1958–1960, 1966

Western wall

120

Brecciated
deposits

**Upper
Cave**

Pigeon Hall Cave

0   10   20   30   M

S

**Lower
Cave**

Walkway entrance

Locality 1

future discoveries would make a fossil Mecca, was then virtually a blank on the map of human fossils. Charles Darwin, intellectual grandfather of the evolutionists, preferred Africa as the source of humanity, whereas Alfred Russel Wallace, codeveloper of the theory of natural selection with Darwin, had postulated Asia as the wellspring of the human lineage. The vast majority of researchers agreed on this point with Wallace. German, Swedish, French, Austrian, and American paleontologists flocked to China for the purpose of finding the evolutionary Garden of Eden, but it was a Swedish geologist, J. Gunnar Andersson, who hit real pay dirt. Andersson discovered, developed, and first brought to international attention the northern Chinese site of Dragon Bone Hill. A quarry known in Chinese as "Longgushan" and located north of the village of Zhoukoudian, it would produce the largest cache of early hominid fossils known up to that time. The massive excavation that uncovered the fossils is today still the largest undertaken at a fossil hominid site. The discoveries at Dragon Bone Hill, more than any other single site, became central elements in the modern interpretation of human evolution.

The Fortuity of Dragons: Longgushan and Traditional Chinese Medicine

A mysterious affinity exists between the ancient dragons of Chinese myth and the fossilized remains of extinct animals. This association was discovered by accident. In 1899 the German naturalist K. A. Haberer traveled to China to explore the natural history of the western parts of the country, but was forced by the Boxer Rebellion to stay on the Chinese coast. In Shanghai, Beijing, and other cities he discovered that Chinese apothecary

---

Facing page

Top: Dragon Bone Hill ("Longgushan") is located 50 kilometers southwest of Beijing, near the town of Zhoukoudian. Located strategically at the point where the Western Hills meet the North Chinese Plain and near the Zhoukou River, Longgushan offered shelter, nearby water, and a vantage point for prey for Pleistocene carnivores, and at times, *Homo erectus. Middle:* The location of ancient dragon bone quarrying was on the northeastern slope of Dragon Bone Hill, but when the site was rediscovered (and renamed "Locality 1") by scientists, excavation began on the northern slope of the hill. *Bottom:* A plan view of Locality 1 with a history of the excavations. The first excavation by Otto Zdansky was in 1921 above what was later named the "Lower Cave" and at the entrance to the site used by visitors today. The last excavation was completed in 1980 under the direction Lanpo Jia. Pigeon Hall Cave ("Gezitang" in Chinese) was originally dug out by generations of dragon bone quarriers. Dragon Bone Hill was designated a United Nations World Heritage site in 1987.

shops sold vertebrate fossils under the names of "long gu" ("dragon bones") and "long ya" ("dragon teeth"). Traditional Chinese believe that the fossilized bones are the remains of dragons—mythical animals associated with rain, clouds, fertility, good fortune, and royal power. Medicine made of ground dragon bones could cure a variety of ills.

Haberer was able to buy quite a few fossils of extinct Chinese animals that, until then, were largely if not entirely unknown to science. Remarkably, included among his collection of "dragon bones" was a molar tooth that was apelike, possibly even human. In 1903 the German anatomist and paleontologist Max Schlosser studied Haberer's collection and published a paper on the finds.[2] In addition to confirming that all of Haberer's dragon bones were in fact mammals, he considered the apelike tooth to be a fossil hominid and the first representative of the long-awaited human precursor from mainland Asia. However, as tantalizing as these fossils were, their provenance—where they came from or how old they might be—was unknown. Organized, scientific fieldwork in China was needed.

Henry Fairfield Osborn, head of the American Museum of Natural History, friend of presidents, and the leading paleontologist of his day, intended to do something about the paleontological void in the Far East. He founded the Central Asiatic Expedition to China in the early 1920s. While visiting the field in 1923, he saw some Chinese peasants pointing at and obviously discussing him and his field director. Asking for a translation, he learned that the Chinese had referred to them as "American men of the dragon bones." Osborn wrote in 1924, "I was delighted with this Chinese christening. For what purpose were we in Mongolia? . . . to collect the bones of dragons—the dragons which for ages past had ruled the sky, the air, the earth, the waters of the earth, and which even today are believed in implicitly by the Chinese."[3] Osborn was so taken with the subject of dragons that he persuaded a colleague to write a book on the subject, to which he penned the introduction.[4] But Osborn's grand plan of finding human "dragon bones" in Mongolia was to fail. Because all the sediments that the American Museum team investigated were far too old for hominids, the years of work yielded not a single scrap of a human ancestor. In keeping with a "gentleman's agreement" to leave scientific exploration in northern China to a remarkable Swede by the name of J. Gunnar Andersson,[5] the American Museum team never went to Dragon Bone Hill.

J. Gunnar Andersson was an explorer, polymath, and scientist who made his living as an economic geologist. He had been head of the Swedish Geological Survey and before that had explored Antarctica. As part of an international effort to map worldwide geological resources he had been seconded by the Swedish government to work for the Chinese Geological Survey, arriving in China in 1914. Andersson's main assignment was to explore the rock units of China in search of economically important resources such as coal, oil, natural gas, and ore-bearing deposits. His publications, however, belie much broader interests. He published observations on Chinese history, archaeological sites, ancient myths, and, most importantly to our story, fossil deposits of paleontological interest. Also an excellent draftsman, he illustrated his books with his own drawings of landscapes and sketches of individuals. When he returned to Sweden in the late 1920s he became the founding director of the Museum of Far Eastern Antiquities, an institution filled with archaeological collections accumulated during his 15 years exploring China (and shared 50–50 with the Chinese government). Andersson was yet another Westerner to come under the spell of the mythical Chinese dragon and its bones. In 1925 he wrote a paper on the archaeo-

Swedish geologist J. Gunnar Andersson worked in China between 1914 and 1926. Following up on a tip by an American chemistry professor, he confirmed the presence of fossil bones near Zhoukoudian in 1918. It was through his continued interest and organizational skills that a program of scientific excavation was begun at Longgushan.

logical history of Chinese dragons[6] and his 1928 memoir of his years in China was entitled *The Dragon and the Foreign Devils*.[7] In his extensive travels around China he paid especial attention to reports of "dragon bones" because, mindful of Haberer's and Schlosser's earlier findings, he knew they could lead to fossil sites.

An American missionary teacher of chemistry in Beijing, J. McGregor Gibb, first told Andersson about some fossil bones that he had seen in the village of Zhoukoudian (then spelled in English as "Choukoutien") in February of 1918. Zhoukoudian, only about 50 kilometers southwest of Beijing, was easy to get to because it was right on the railroad line. Gibb had even collected some of the bones and showed them to Andersson. The small fragmented bones were white and fossilized, and they were covered with a red clay that Andersson recognized as a common type of cave sediment in northern China. Andersson, already in China for four years, was excited that this site might actually be one of the sources of the apothecaries' dragon bones.

On March 22 and 23 of 1918, the soonest he could arrange it, Andersson visited the village of Zhoukoudian. Locals took him to an outcropping of red clay-like rock standing as an isolated pillar in the middle of an old limestone quarry. Much of the limestone from which the buildings of Beijing were built came from Zhoukoudian quarries such as this. Andersson saw many small bones protruding from the sediment. The translator told him that this place was known as Ji Gu Shan (now written "Chikushan") or "Chicken Bone Hill." Locals took the small bones to be those of animals with which they were familiar—chickens. Andersson, however, recognized most of them to be rodents' bones, and, in one instance, a large mammal bone. He excitedly wrote down the location of the deposit of bones and his observation that the area had potential paleontological importance. He was curious as to why the quarrymen had left the deposit of bones when it would certainly have been less trouble to simply dig through it into the limestone. His question was answered by the villagers: "Once upon a time, more than a hundred years ago, there was a cave here in which lived foxes, which devoured all the chickens in the neighborhood. In the course of time some of these foxes were transformed into evil spirits. One man tried to kill the foxes, but the evil spirits drove him mad."[8] Andersson then understood not only why the pillar was left standing, but why the villagers had had no hesitation in showing him and other foreigners the enchanted fossil deposit. But madness or no, Andersson determined to come back to this place.

Back in Beijing, Andersson's other projects intervened, and it was three years later, 1921, before he was finally able to return to Zhoukoudian and Chicken Bone Hill. This time he came with a paleontological assistant who was a recent student of Swedish professor Carl Wiman, named Dr. Otto Zdansky, originally from Vienna. Andersson had brought Zdansky to China mainly to excavate rich deposits of three-toed horses (*Hipparion*) that he had discovered in Henan Province, and Chicken Bone Hill was to be a practice run. When the eminent American paleontologist, Dr. Walter Granger, the first of Osborn's American Museum team to arrive in China, showed up, Andersson invited him to come along to visit Zdansky in the field. Andersson thought that Granger could give Zdansky some useful tips on the latest American excavation methods.

When Andersson and Granger arrived at Zhoukoudian, Zdansky had set up camp in the local temple and was at work at the site. All three set to work on digging out, preparing, and labeling the fossils coming out of the site. While the three scientists were at Chicken Bone Hill, a man from the town came out to see them. After watching for a while, he said, "There's no use in staying here any longer. Not far from here there is a place where you can collect much larger and better dragons' bones."[9] The villagers had probably been thinking of how best to get rid of these foreigners, especially the one camped out long-term in their temple. Information about a valuable dragon bone locality might lure the Westerners away from town. It worked.

The man led Andersson, Zdansky, and Granger north, across the foot-bridge over the river, out of town, past the railway station, and up into the limestone hills. The villagers watched them go, carrying their excavation equipment with them. About 150 meters above the station they came to an old abandoned quarry, which had also been mined for building stone. It faced northeast, diagonally away from the town. Here the man showed Andersson and his colleagues a fissure in the limestone cliff face filled with fossil bones. Within the hour they had found the jaw of an extinct pig. It was clear that they now had a site with much greater potential than Chicken Bone Hill, and they decided to move operations immediately. Andersson wrote, "That evening we went home with rosy dreams of great discoveries."[10] When the man returned and informed the townspeople of Zhoukoudian of the developments, it is more than likely that they were pleased as well.

Early the next morning, Andersson, Zdansky, and Granger walked from the temple to the new site. What they found "exceeded all expectations." They discovered fossil jaws of the extinct giant elk, later to be named *Megalotragus pachyosteus*; hyenas; bears; and many other fossils. Granger showed Zdansky how to apply supporting plaster jackets to the fossils—the method the Americans had developed to preserve fossils in the field. After one full day at the site, Andersson concluded that Zdansky had weeks of work ahead of him just jacketing, preparing, and recording the fossils. Andersson and Granger planned to take the train back to Beijing the next day.

The new site was initially referred to as "Lao Niu Gou," which trans-lates as "Ravine of Old Niu" (Niu being a surname). When Zdansky pub-lished his report,[11] he named the site after the nearby town, "Choukoutien" (Zhoukoudian, or "shop on the Zhoukou [River]"), by which name it has gone in scientific circles ever since. But the Chinese call the site "Longgushan"—Dragon Bone Hill. J. Gunnar Andersson and Otto Zdan-sky are given credit for the scientific discovery of the Longgushan fossil site above the village of Zhoukoudian. But, in truth, this deposit of large and hardened "dragon bones" had been known to local Chinese for centuries. The name of the Zhoukoudian townsperson who led the Westerners to their "discovery" has been lost to us.

What went on behind the scenes in Zhoukoudian to set in motion the discovery by Western science of Longgushan has also not been recorded. We may presume that the Zhoukoudian dragon bone diggers, whose occupa-tion passed from father to son, were either willing to transfer operations to another quarry site of which they knew, or simply had their protests drowned out by the townspeople who wanted a solution to the presence of foreign devils in their temple. It is more than likely that the original Zhoukoudian dragon bone diggers saw more potential profit working for the scientific excavators at Zhoukoudian than digging for the bones themselves. Zdansky hired some dozen men to assist him in the excavations.

There was certainly also in Zhoukoudian a sizable number of townspeople who believed that the Westerners had desecrated the temple (now also used as a local school) and should be driven from the land altogether. After all, this had been the feeling of many Chinese people during the so-called Boxer Rebellion of 1899–1900, a popular uprising against foreigners in China brought on by the occupation of Chinese territory for economic gain by German, French, Japanese, and British forces. A similar popular protest would occur during the 1925 Shanghai massacre of Chinese students by foreign policemen. For the traditionalists, it was for the dragon, the protector of the land and bringer of rains, to dispense with the foreigners.

Indeed, the very morning after the initial exciting discoveries at Longgushan, great clouds covered the sky and then unleashed torrential rains. The little Zhoukou River flowing through town overflowed its banks and washed away the bridge, cutting the scientists off from their new site. Andersson and Granger could not get to the railway station. Andersson relates that he and Granger "were hopelessly flooded in, for the little stream which flows out into the Chou K'ou Tien valley, and which during the preceding days had been an insignificant purling rill, was now a wild foaming mountain stream that nobody dared to cross so long as the cloudbursts continued to hurl new masses of water into the valleys."[12] For three days the scientists huddled in the temple, telling stories and drinking, until the rains let up. To escape Zhoukoudian on the fourth day, Andersson and Granger had to wade across the river "almost naked," holding their clothes and shoes above their heads, undoubtedly to the twitters of many townspeople. Some saw in these events the power of the dragon, which had stopped the foreigners in their tracks and had made them retreat ignominiously from Zhoukoudian.

Suspicions of Hominids at Longgushan and Their Discovery

One of the foreign scientists remained, however. The stubborn young Austrian, Otto Zdansky, continued working at Longgushan for another four months, until the end of the summer of 1921. He worked on the baking hot limestone cliff face with his field laborers extracting bones, cleaning off the adhering sediment, gluing broken pieces back together, putting plaster jackets on the larger pieces, and recording everything. When Zdansky finished his work at Zhoukoudian, the fossils he had collected were shipped via Beijing to the laboratory of Professor Wiman in Uppsala. Meanwhile, he went off to Henan Province to undertake the main task for which Andersson had brought him to China—to excavate three-toed horses.

Zdansky advanced several reasons to Andersson for leaving the work at Zhoukoudian. From a purely paleontological standpoint, the fossil specimens at the site were fragmentary and not extremely well-preserved. The sediments enclosing the fossil bones were very hard, and they tended to break along lines that fragmented the fossils. Finally, as Zdansky and his workers had quarried into the cliff face, an overhang had formed, looming dangerously over their heads. Andersson acquiesced, and Zdansky moved on to his next challenge, in southern China.

Andersson did not forget about Zhoukoudian. On one of his visits to check up on Zdansky during the summer of 1921 he had paid particular attention to angular pieces of quartz that were associated with fossil bones and that were found in two layers of the deposit. One of Andersson's abiding interests in China was archaeology, and he immediately seized on these quartz flakes as possible stone tools of fossil hominids. Zdansky pointed out that there were plenty of quartz veins within the limestone from which the fragments could have naturally derived. Andersson had to admit that natural erosion from the roof or walls of the cave was "the most probable, or at any rate the least sensational, interpretation of the occurrence of the flakes of quartz."[13] But not to be dissuaded, he postulated that the earliest hominids, before they actually fashioned stone tools, picked up naturally occurring stone and wood for tools. One day at the site, Andersson knocked on the side of the limestone wall and prophesied, "I have a feeling that there lie here the remains of one of our ancestors and it is only a question of your finding him. Take your time and stick to it till the cave is emptied, if need be." But the quest for early hominids was Andersson's fascination, not Zdansky's, and as we have seen, Zdansky had advanced good reasons for discontinuing the work at Longgushan.

In the summer of 1923, after Zdansky had had considerable success in excavating the *Hipparion* sites and in discovering numerous other paleontological riches (he later had new species of a sauropod dinosaur and a fish named after him), Andersson succeeded in persuading him to return to Longgushan. The year before, Zdansky had constructed scaffolding and skillfully extricated a huge block of mammal fossils from a vertical cliff wall in Kansu Province, and thus he could no longer use this excuse for refusing to return to Zhoukoudian. Zdansky is reported to have said, "I wasn't interested in what Andersson wanted. I wanted only the fauna of the cave."[14] He undoubtedly went back to the site he termed Zhoukoudian because he wanted to bolster the research paper that he would write about its paleontology, but Zdansky also had a secret that he knew made the site immensely more important.

Some time late in the summer of 1921 Otto Zdansky discovered a single molar tooth of what he identified in the field as an "anthropoid ape." He recognized it as the long-sought-after hominid but, remarkably, he did not tell Andersson. Speaking to journalist John Reader 57 years later in Uppsala, Sweden, Zdansky said, "I recognized it at once, but I said nothing. You see hominid material is always in the limelight and I was afraid that if it came out there would be such a stir, and I would be forced to hand over material I had a promise to publish."[15] Reader also reported that Zdansky harbored ill will toward Andersson after an initial argument the two had had soon after Zdansky's arrival in China. In 1923 Zdansky sailed for Sweden, taking the fossils from Longgushan, and the newly discovered hominid tooth, with him. Andersson was not to know of the discovery until 1926. Back at Professor Carl Wiman's laboratory at the University of Uppsala, Zdansky had time to clean, catalog, and study the fossils he and his excavators had extracted from Longgushan. He could mull over his hominid molar and carefully frame and articulate his conclusions.

Finding only one specimen of a previously unknown species is always a quandary for a paleontologist. The questions abound. Is it really a record of a new species, or could it be something else, say a fragment of another animal species just masquerading as a new species? Or could it be a skeletal element from a later time that somehow became incorporated into the fossil deposit? This was not an unreasonable thought for a potentially human fossil that could have been buried by human hands much later than the other fossils had been deposited. Even if it was a higher primate molar tooth, how sure was he that it was not some kind of ape or monkey? Zdansky was cautious. He was a young and inexperienced Ph.D., just starting out, and he knew that whatever he said about the fossil anthropoid molar, actually a very minor part of the overall Longgushan fossil assemblage, might well overshadow all his other work.

Two discoveries that Zdansky made in the laboratory in Uppsala helped him make a decision. First, he discovered among the many isolated teeth from the excavations a few isolated teeth of a previously unknown fossil monkey, a macaque. The hominoid (that is, apelike or human) molar looked nothing like the monkey teeth. Then, some time in 1924 or 1925, he found a second fossil hominoid tooth—a premolar. The premolar had a low and flattened crown like a human, and very unlike an ape. With two fossil teeth now in hand, and a clear argument that they did not represent a previously unknown monkey or ape species, Zdansky felt confident in reporting to Professor Wiman that he had a fossil hominid among the Longgushan fossils. Still, he downplayed their importance, referring them

The first two hominid teeth found by Otto Zdansky at Longgushan in 1921–1923 were informally termed "*Homo pekinesis*" by Davidson Black, and popularly dubbed "Peking Man." They are shown together with a third tooth found later among the Longgushan faunal collections housed in Sweden. All of the collections made by Zdansky at Longgushan from 1921 to 1923 still reside at the Paleontological Institute at Uppsala.

to the conservative (and, by the way, still accurate) taxonomic category of "*Homo* sp. ?"[16]

   J. Gunnar Andersson received the new information about the discovery of hominids at Longgushan, not by Zdansky but by his professor, Carl Wiman, in a letter sent in mid-1926 from Uppsala to Beijing. Andersson had requested from Wiman an update on the paleontological collections that had been sent back to Sweden for identification. Amazing discoveries had been made—a new Chinese dinosaur, unusual fossil giraffes, and a unique species of long-snouted, three-toed horse. But Andersson honed in on Zdansky's report on the small and fragmentary teeth from Longgushan, exclaiming excitedly, "So the hominid expected by me was found."[17] Andersson had been kept in the dark for five years by Zdansky's secrecy.

The American "Missing Link Expedition" Goes on a Wild Dragon Hunt into the Gobi

Paleontologist Walter Granger of the American Museum of Natural History had been in on Andersson and Zdansky's discovery of the Longgushan site in 1921. News of this discovery was added to Granger's report back to New York to museum director and paleontological czar, Henry Fairfield Osborn. Osborn, friend of Teddy Roosevelt and president of the American Association for the Advancement of Science, sat in his large leather chair behind his massive desk in one of the four towers of the castellated edifice that he had done so much to build, and pondered China. Such was Osborn's power in international scientific circles that it never occurred to him that even if he decided that he wanted Longgushan, he could not have it. A "missing link expedition" through China to the Gobi Desert would ultimately result.

Osborn's calculations involved many factors. Was there a strong scientific presence there already? No. Andersson was only an economic geologist who needed Wiman and Zdansky to identify his fossils. Even if Andersson felt some ownership of the site, thought Osborn, he was not going to be able to investigate it himself. Osborn knew from his contacts in Europe that Zdansky did not like Andersson and did not want to return to China. Zdansky was also in line for a job at the University of Cairo in Egypt, which he eventually took. No, Zdansky would not stand in his way.

What was the museological value of an excavation at Longgushan? Immense, thought Osborn. There was tremendous interest among the public in the evolutionary link between man and the lower primates, and if Osborn could put such a fossil on display in his museum, the public would flock to it. Osborn had himself predicted Asia would be the place to find such ancestors. And he had the world's foremost array of technicians, scientific artists, and associate scientists to collaborate in the ensuing publications.

Was it feasible? Osborn had done the bold and unthinkable before. Desiring missing links and complete skeletons to fill his hall of elephants, he had dispatched well-equipped teams from New York to the fossil badlands of the American West. No expense had been spared, and the skeletons had been found, studied, mounted, published, and finally exhibited—to universal acclaim. Mounting an expedition to Asia to find the human missing link would be even more challenging. Could he do it? Osborn decided that he could, and the Central Asiatic Expedition, perhaps the most lavishly funded and massively organized effort ever mounted to find fossil hominids, was born. The expedition began its work during the summer of 1922, with Osborn's handpicked successor, Roy Chapman Andrews, in charge. Osborn intended the discovery of ancient human ancestors in Asia to be his swan song, the most dramatic culmination of an impressive career. J. Gunnar Andersson, however, had other ideas.

After receiving Wiman's letter with Zdansky's news about the two hominid teeth from Longgushan, Andersson starting making his own endgame plans to discover early hominids in China. He drew on a number of resources unknown and unavailable to Henry Fairfield Osborn. First of all, Andersson was setting the stage for his own departure from China. He had long ago made a contract with the Chinese government to share fossil and archaeological collections between China and a new museum that he was planning back in Stockholm, the Museum of Far Eastern Antiquities. Andersson had been quietly and systematically collecting for this purpose for 15 years in his extensive travels around China, and he was to be founding director of the new museum.

Over the years Andersson had built a reputation as a trustworthy and honorable man in his dealings with both the Chinese and Westerners in China. His knowledge of the country, its sites, and its people was virtually unparalleled. Andersson was in a position to know where and with whom to throw his lot. With the exception of Granger, Osborn's people were new to China and were at a distinct disadvantage in knowing the lay of the land.

Andersson was also not quite the simple economic geologist that Osborn and perhaps others imagined. Behind Andersson's work in China was substantial financial backing. An influential benefactor back in Sweden, industrialist Dr. Axel Lagrelius, had set up and endowed a foundation called the Swedish China Research Committee. Lagrelius was a friend of the crown prince of Sweden (later King Gustavus VI), who agreed to serve as Chairman of the Swedish China Research Committee. It had been funds from this source that had paid the salaries of the Longgushan excavators and Zdansky, paid for the shipments of fossils from China, and helped pay for ongoing expenses at Wiman's laboratory.

As luck would have it, the crown prince was to arrive in Beijing on an around-the-world tour in October 1926. Dr. Lagrelius traveled to Beijing to be there when the prince arrived. Andersson found himself in charge of arranging events for the prince's "archaeological and art studies." By the time the prince arrived, Andersson and Lagrelius had laid careful plans and skillfully engineered a scientific meeting and social event so influential that it was to block any hopes Osborn may have had for his Central Asiatic Expedition ever excavating at Longgushan. The meeting would launch the name of "Peking Man" and set in motion the series of hominid discoveries for which the site near Zhoukoudian would become world famous. It also represented a cementing of scientific alliances across international boundaries and brought into Andersson's circle the influential and American-funded Peking Union Medical College. Osborn's grand vision disappeared in a cloud of Gobi Desert dust, as his expedition toiled hundreds of miles and millions of years distant from the true early hominids of ancient China.

John D. Rockefeller's Chinese Medical School and Its Unruly Anatomist, Davidson Black

China in the early twentieth century was a country in economic and political chaos. The country's vastness and economic importance had prompted the imperial powers to take control of parts of the country, particularly the ports, after the Boxer Rebellion, but at the same time, Westerners and their institutions became involved in a variety of humanitarian causes in China. One large American foundation, the John D. Rockefeller Foundation, acted to fund the establishment of an English-language medical school, the purpose of which was to train young Chinese doctors. With excellent salaries, the Peking Union Medical College was staffed by adventurous faculty from all over the world, but there was a preponderance of North American professors.

Davidson Black was hired by Peking Union Medical College in 1919 as professor of anatomy.[18] A Canadian, Black had an M.D. from the University of Toronto, but after a short stint in World War I, he had spent time traveling to the laboratories of prominent physical anthropologists in the United States, England, France, Holland, and Germany in order to learn as much as possible about human evolution. He had found a mentor in Dr. (later Sir) Grafton Elliot Smith, eminent professor of anatomy at University College in London. As his letters indicate, Black was interested in the China job, mainly because he would be near sites that he suspected might contain fossils of human ancestors. It is almost certain that the officers of the Rockefeller Foundation who decided to hire young Davidson Black, M.D. for the position in anatomy at Beijing had no idea that his anatomical research would involve digging for fossil bones in an old, dusty stone quarry many miles and many hundreds of thousand of years removed from the newly built medical school in Beijing. Elliot Smith wrote Black a sterling recommendation for the job.

In the first two years that Black was in Beijing he threw himself into the job of organizing and building up the medical school, particularly the anatomy department. His anatomy lectures went well, and he developed good relationships with his colleagues. One of his jobs was to obtain ca-

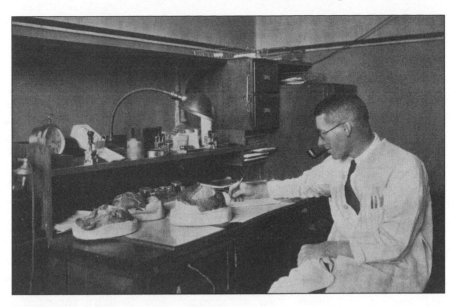

Davidson Black at his laboratory workbench in the anatomy department of Peking Union Medical College with *Sinanthropus* skulls. It was here that Davidson Black died around midnight on the evening of March 15, 1934, flanked by Skull III and the Upper Cave skull.

davers for the medical students' dissection. The Beijing police were only too happy to oblige and one day sent over to the anatomy department a number of headless corpses of executed criminals. Shocked, but always diplomatic, Black visited the police and explained that he needed intact bodies for the medical school. The police chief listened and then nodded. Some days later a line of shackled prisoners arrived at Black's office from the police station with a note saying "kill them any way you like." This turn of events, of course, occasioned another trip by Black to the police station.[19]

Black had an engaging and outgoing personality, and he and his wife were active in the social life of expatriate Beijing. He also kept up a lively correspondence with his friends and colleagues abroad. The Peking Union Medical College was well pleased with Black, and he was appointed chairman of the anatomy department.

In 1921 Davidson Black began a collaboration with J. Gunnar Andersson at the Neolithic cave site of Shaguotun, northwest of Beijing in Manchuria. The two men had undoubtedly met on social occasions before this, because in describing their first work together Andersson calls Black "my friend."[20] Andersson had been working in Manchuria assessing coal resources, but he set his assistants to excavating the interesting Shaguotun cave nearby. Returning from the coal deposits, Andersson was pleased to find that they had discovered a large number of human bones. Andersson immediately wired Black for help in the excavation and anatomical study of the human skeletons. They were only a few thousand years old (versus several hundred thousand years for the Longgushan fossils), but Black was still interested. He arrived by train at the site on June 22, 1921. The bones went back to Black's lab at the medical school where they were cleaned and studied. Black found that the bones had come from some 45 individuals, but their remains had been jumbled, broken, and probably cannibalized. He eventually published his results in the Chinese journal that Andersson had helped found, *Palaeontologia Sinica*.[21]

The Peking Union Medical College administration was not pleased with Black's newly evinced interest in physical anthropology. Dr. Henry Houghton, president of the college, told him in no uncertain terms to limit his research to medical subjects, not "mythological caves." Houghton, an M.D. trained at Johns Hopkins University, knew little about physical anthropology and its close relationship to anatomy. Unlike at most European universities and medical schools, where physical anthropology had been an established part of the curriculum for two generations, in the United States formation of the American Association of Physical Anthropologists was still in the future. (It was founded in 1930.) If Black had not been so competent in all other realms and so universally well liked, the medical school administrators would probably have found a way to rid themselves of this budding paleoanthropologist. After the skeletal remains from Shaguotun had arrived in Black's lab, he was able to strike a deal with the foundation. He agreed to delay the research on the bones for two years, during which he would spent his days teaching in the medical school and working on anatomy department business. In 1922 Black also turned down an offer from Roy Chapman Andrews to work as an anatomist for the American Museum's "Missing Link Expedition," either as part of his agreement with the medical school or because he had already established a firm working relationship with Andersson the year before.

Even after the two years were up, however, Black found that he still faced administrative objections to his paleoanthropological activities. For example, Dr. Houghton refused to pay for an invited lecture in Beijing when he learned that the lecturer was to be a well-known physical anthropologist, Aleš Hrdlička, of the Smithsonian Institution. Eventually, the Rockefeller Foundation back in New York made a contribution to the Smithsonian to cover the cost of Hrdlička's travel. It was clear that Davidson Black had a problem, and one wonders if his fabled propensity for working on his research in the dead of night originated from his desire to keep a low profile and to avoid confrontation with medical school administrators, all of whom could be relied upon to be home in bed when Black was working on his skulls and bones.

By 1926 Black had published the results of his analyses of the Shaguotun remains. When Andersson asked him to participate in the scientific meeting planned for the Swedish crown prince, Black agreed, but he realized that Andersson could also help him. Andersson and Black were clearly in cahoots in organizing the media event that occurred on October 22, 1926.

Andersson handed the special lantern slides that Zdansky and Wiman had made of the two hominid teeth in Uppsala over to Davidson Black. Black worked up a short description of the teeth for Andersson to present at the meeting, and then sent the paper off to the journal *Nature*, which published it a month later.[22] The meeting itself started with talks by the Chinese head of the Geological Society, Weng Wen-hao, a Chinese political reformer, the French Jesuit paleoanthropologist Pierre Teilhard de Chardin, and finally Andersson, who reported on Wiman's paleontological research and his own archaeological finds. Last came the coup de grâce, the lantern slides of the Longgushan hominids.

Andersson, with feigned indifference, concluded that he had no plans to pursue these remarkable discoveries, but it would be a shame not to follow them up. He proposed that Peking Union Medical School, whose representative, Dr. Black, was at the meeting, and the Geological Survey of China, headed by Andersson's long-time friend and colleague Dr. Weng Wen-hao, collaborate to mount such a project. It was a daring move, and because of the circumstances of the meeting—it was virtually a royal hearing—all eyes turned to the crown prince for a response. The prince, an amateur archaeologist himself and intimately informed of Andersson's untiring efforts over the last 15 years, gave his enthusiastic support. Andersson, for his part, needed the prince's backing to continue legislative and funding initiatives back in Sweden to get his Museum of Far Eastern Antiquities off the ground, but the prince had no difficulty supporting Andersson's suggestion for continuing work in China. After all, he was not being asked to fund it (although it was clear to everyone at the meeting that the Swedish China Research Committee, which the prince chaired, had paid the way up to that point). The prince was also impressed with Andersson's marshaling of the scientific results from these logistically complicated and long-term explorations. Andersson's international stature was confirmed by the show of support from the obviously very capable Canadian anatomist; by the backing of both the Chinese scientific establishment and the progressive political elements in China; and by the full participation of the eminent French paleontologist, Teilhard de Chardin. Andersson got what he wanted out of the meeting. It was a grand send-off from China for him.

Davidson Black also got what he wanted out of the meeting. In 1926 he had been at the Peking Union Medical School for seven years, and during that time he had done an excellent job but had shown no indication that he intended to stop anthropological research. So it was perhaps time that the Rockefeller Foundation made peace with Black and his anthropological interests. The visibility of the meeting with the Swedish crown prince; the publication in *Nature*, the first by any faculty member at the Peking Union Medical College; and the broad international acclaim for the importance of the new site near Zhoukoudian all combined to bring the Rockefeller Foundation around to Black's point of view. The foundation agreed to fund the formation of a "Cenozoic Research Laboratory" at the Peking Union Medical College, with Davidson Black as honorary director, and to provide funding for the excavation of Longgushan. This three-institutional collaboration of the China Geological Society, the Peking Union Medical College, and the Rockefeller Foundation was to continue at Longgushan until World War II eventually halted the research nine years later.

The meeting in Beijing for the crown prince of Sweden bequeathed one more lasting legacy to paleoanthropology. In the press coverage resulting from the meeting, the term "Peking Man" was born. In an interview immediately after the meeting, Dr. Amadeus W. Grabau, a German-American invertebrate paleontologist and professor of geology at Beijing University, was quoted as using the name to refer to the two fossil teeth discovered by Zdansky. This is the colloquial name by which the fossil hominids from Longgushan near Zhoukoudian have been known ever since. Grabau was also a close friend of Andersson, who includes a sketch of him on page one of his book *Children of the Yellow Earth* and describes him as "a scholar of genius, an enthusiastic teacher, and a delightful man."[23]

The birth of "Peking Man" was not to be without incident. The worst fears that Zdansky had harbored regarding any identifications of hominid remains from Longgushan came to pass. Someone, and not just anyone, questioned the identification. None other than Professor Father Pierre Teilhard de Chardin wrote a letter to Andersson two days after the meeting with the prince. It was brief and to the point. In regard to the two fossil teeth from Zhoukoudian, he was "not convinced of their supposed human character," instead suggesting that both specimens might be the worn or fragmentary back teeth of carnivores. He did note that he had not examined the original specimens, only Andersson's photos, and that he hoped "intensely that my criticism will prove unfounded."[24]

Teilhard's criticism shot around the Beijing scientific community like an electric shock. Teilhard and his French archaeologist colleague, Emile Licent, who had also been at the meeting, had clearly not been in on

Andersson's and Black's plans. Perhaps Teilhard felt a little put out by being excluded, or perhaps he simply felt that the teeth were not hominid and that it was his duty to communicate this opinion to Andersson. In any event, the doubt over the identity of the two teeth from Longgushan threatened the whole enterprise, but particularly the reputation of Davidson Black, who had a paper in press in the most prestigious scientific journal in the world at that very moment, supporting Zdansky's identifications. But if Black was worried, he made no great show of it. And when Grabau ribbed Andersson in front of Teilhard and some visiting French scientists about whether Peking Man was a man or a carnivore, Andersson replied, for no apparent reason other than to have a quick comeback, that it was neither, but a lady.[25] The jocularity helped, but a pall was to hang over Peking Man and Davidson Black until the site could produce definitive hominid remains that could silence the skeptics.

The Coming of *Sinanthropus*

The Rockefeller Foundation for its part backed Black, and funding for the joint excavation at Longgushan went forward. But the Swedes were not quite out of the equation yet. The coalition asked Andersson and Wiman to help organize the excavation. The Rockefeller Foundation was particularly concerned that Black not be taken away from his duties at the medical school. By this time Otto Zdansky had published his paper on the initial results of the site and had no interest in coming back to China. Another one of Wiman's students, Dr. Birger Bohlin, who had studied some of the fossil giraffes from China, came out in 1927 to oversee the excavations at Longgushan.

Bohlin was a young and enthusiastic fieldworker. He had sailed to China with his wife, who lived in Beijing while he was at Longgushan. Had he been older and more experienced, he might well have been much more apprehensive of the situation into which he was headed. China was still occupied by very unpopular British, German, French, Japanese, and American military contingents, protected by the "unequal treaties" militarily forced upon China. The first president of China, nationalist Sun Yat-sen, who had been elected in 1912 after the collapse of the 268-year-old Qing Dynasty, had died in 1925. In 1927 the Nationalist party under Chiang Kai-shek, the Chinese Communist party with its future chairman Mao Tse-tung, and various local warlords were all vying for territory, power, and supremacy in China. The armies of two feuding warlords, Chang Tso-lin and Yen Hsi-shan, were fighting within earshot of the town of Zhoukoudian. Bohlin frequently saw troops marching back and forth, heard the cannon fire of their battles in the distance, and occasionally had to deal with bandits passing the excavation. But miraculously, no major incident marred his fieldwork. He started excavating on April 16, 1927, and finished only when he had discovered the long-awaited hominid, six months to the day after he began, on October 16. In all, Bohlin and his team moved three thousand cubic meters of cave sediment.

The fossil hominid that Bohlin found was only one tooth. But that did not stop his being elated at the discovery. As soon as he had closed down his operations at Longgushan, he hurried back to Beijing, avoiding soldiers and bandits along the way. He arrived at Davidson Black's lab at 6:30 P.M. on October 19, before he had even cleaned up or told his wife that he was back in Beijing. Black described him as "covered with dust but beaming with pleasure."[26] When he saw the tooth, which was well preserved and undoubtedly hominid, Black was overjoyed. It had been a year since the meeting with the prince, and six long months of excavation. Bohlin had shipped back from the field a large number of wooden crates of fossil-containing sediment to be prepared in Beijing. Black noted that "Bohlin is quite certain that he will find more of *Homo pekinensis*."

What Davidson Black did next has been considered remarkably prescient, politically expedient, or foolhardy and irresponsible, depending on one's perspective. On the basis of the single tooth that Bohlin had brought back to him, Black named a new genus and species of hominid, *Sinanthropus pekinensis*, published in *Palaeontologia Sinica* within a few weeks of discovery.[27] It would perhaps have been more responsible to have at least waited until

GEOLOGICAL SURVEY OF CHINA

V. K. TING AND W. H. WONG, DIRECTORS

Palæontologia Sinica

EDITORS:

V. K. TING AND W. H. WONG

Series D. Volume VII.
Fascicle 1.

THE LOWER MOLAR HOMINID TOOTH FROM THE

CHOU KOU TIEN DEPOSIT

BY

DAVIDSON BLACK

PLATES I-II AND TEXT FIGURES 1-8

PEKING 1927.

The first human fossil discovered at Longgushan when excavations were resumed in 1927 was a lower molar tooth that anatomist Davidson Black named a new genus and species, *Sinanthropus pekinensis*. This is the cover page of Black's 1927 paper announcing the new species.

Bohlin had had time to prepare the other hominid fossils that turned out to be in the sedimentary matrix, but Black decided to move fast. He did so, undoubtedly, to dispel the year's worth of doubt over the reality of Peking Man, and because more funding was needed from the Rockefeller Foundation to continue the excavations next season. With a formal Latin name and confidence that more fossils were on the way, Black sailed for North America and Europe. His mission was to lobby for acceptance of his new taxonomic name and to gain new funding from the Rockefeller Foundation. By the time he had returned to China, his diplomacy and persuasiveness had paid off. Funds were granted to continue the excavations. And *Sinanthropus*, based on a single tooth, was accorded more general scientific acceptance than another new hominid genus and species, *Australopithecus africanus*, based on an entire fossil skull and mandible with complete dentition, published two years previously[28] by another Elliot-Smith-trained anatomist, Raymond Dart.

Wenzhong Pei Discovers the First Hominid Skull

Davidson Black had dug himself a scientific hole as deep as the exacavations at the Zhoukoudian cave. To avoid its becoming a professional grave, he needed more fossils. A skull would be critical for the eventual acceptance of *Sinanthropus pekinensis,* simply because so much of the identity of a mammal species is evinced by its facial, cerebral, ocular, nasal, and dental anatomy. The first fragmentary skulls of Peking Man were finally discovered in 1928.

After a winter of lab work, Birger Bohlin was ready to get back to the field in the spring of 1928. Dr. Zhongjian Yang (whose name was then anglicized as C. C. Young), a newly graduated paleontologist (China's first) trained at the University of Munich at the suggestion of Professor Grabau, and Wenzhong Pei (later Dr., then rendered as W. C. Pei), another of Grabau's students, were to be assisting this season. Upward of 60 workers were to be hired at the site.

The excavations of 1928 started near the point at which the fossil molar had been found the previous year, in the northwestern part of the cave. About ten meters higher in the section of the limestone, more teeth, fragments of mandibles, and pieces of skull were found. Bohlin wrote to Andersson back in Sweden about this "whole nest of *Sinanthropus* remains."[29] Back in the Beijing lab, sediment from the original molar site was being slowly broken down and prepared, and more *Sinanthropus* teeth and bones were also being found, as Bohlin had predicted. The researchers called the "nests" of fossils "loci." "Locus A" was the 1927 point of discovery of the first molar and "Locus B" was the new cluster of hominid fossils.

Lower jaws were the first fossils of Peking Man's head to be found. The sediment from Locus A gave up a right half of an adult mandible, and Locus B revealed a juvenile jawbone with its chin region intact. Davidson Black, with characteristic alacrity, published his descriptions of the specimens early the next year.[30] His conclusions are quite interesting, not so much for what they indicate about the ultimate identity of the species *Homo erectus*, but how much they reveal about what Davidson Black expected the new species to look like.

Black's 1929 paper, published in the *Bulletin of the Geological Society of China*, emphasized how apelike the profile of the Locus B juvenile *Sinanthropus* chin region was. It looks in his figures as close to the same angle as the chin region of a young chimpanzee, which he illustrated, and very distinct from the jutting, pointed chin of a modern Chinese child. As an implied evolutionary progression, he illustrated a Late Stone Age Chinese jaw, between the Longgushan specimen and the modern human jaw.

Black wondered what sort of skull went with this apelike jaw. Only a few skull bones, as yet imperfectly cleaned in 1928 (Skulls I and II from Locus B), gave him some idea of cranial form. The bones were quite thick but they were fragmentary. Andersson[31] summarized Black's conclusions: "The *Sinanthropus* corresponds very closely with modern man in size of brain." In retrospect, this is a very surprising deduction for a species now known to have an average brain size only three-quarters the size of modern *Homo sapiens* brains. What could Black have been thinking? Almost certainly, Black's initial conception of Peking Man was that of *Eoanthropus dawsoni*—Piltdown Man—the fraudulent English chimera of ape jaw and modern human skull that masqueraded as a hominid ancestor until 1953.[32] He knew that his knowledge of the true skull form of *Sinanthropus* was very imperfect and that an intact skull needed to be discovered at Longgushan for the mystery to be resolved. That *Sinanthropus* would turn out to be a very different animal from Piltdown was to be Davidson Black's biggest shock.

The 1929 field season saw a changing of the guard at the Longgushan cave site. The last of the Swedish contingent, under whose guidance Longgushan had progressed from enchanted dragon bone quarry to world-renowned hominid fossil site, now left the fieldwork in the able hands of Dr. Yang and Mr. Pei. Birger Bohlin joined another Swedish field expedition to western China after he had finished the 1928 field season, and then ultimately returned to Sweden. The excavation of Longgushan was henceforth to be a Chinese undertaking. The work continued with renewed vigor beginning in April 1929.

Yang and Pei expanded the excavation program. The fossils poured out of the old cave site, and although most were broken and fragmentary, there

The research team in the village of Zhoukoudian in 1929. *From left to right:* archaeologist Wenzhong Pei, who later in the field season would discover the first intact skull of Peking Man in the Lower Cave; field assistants Hengsheng Wang and Gongmu Wang; paleontologist Zhongjian Yang, as the first Chinese excavation head for the project, he published extensively on the fossil vertebrates from the site; Swedish paleontologist Birger Bohlin, who directed excavations in 1927 and 1928; Canadian anatomist Davidson Black, indefatigable professor of anatomy at Peking Union Medical College, and first honorary director of the Cenozoic Research Laboratory in Beijing where the fossils were studied; French Jesuit cleric Pierre Teilhard de Chardin, influential Pleistocene geologist who studied many aspects of Dragon Bone Hill geology, paleontology, and archaeology; Irish geologist George Barbour (later of the Cincinnati Museum of Natural History) who studied geology of the site.

were many beautifully preserved specimens. A complete and intact skeleton of the giant Pleistocene hyena, now known as *Pachycrocuta brevirostris*, was discovered. The list of species of animals discovered at Longgushan continued to grow. More fossils of a non-hominid primate, an extinct macaque monkey, were found. But throughout the long months of digging, the hominid skull that would nail down the identity of Peking Man eluded the excavators.

There is an old superstition that is virtually universal among veteran paleoanthropologists—you always discover the best fossils at the very end of the field season. This had happened at the end of the 1927 season at Longgushan when Bohlin had found the first tooth, and it happened again in 1929.

December had come to northern China and the first snows had fallen in the hills surrounding Zhoukoudian. Pails of water froze overnight. The excavation at the cave had followed a fissure filling, replete with fossils, into the depths of the cave. Only three men could fit at the bottom of this narrow, dark, and cold hole, dug down into what was termed the "Lower Cave." They dug by candlelight. They had dug longer than normal because Wenzhong Pei had a hunch they would find something important.

On the afternoon of December 2, Pei's rock pick pulled away a piece of consolidated sandy and pebbly cave sediment that revealed a tantalizingly interesting round surface of bone. His heart jumped as he carefully began to clean off the edges of the fossil, which was still embedded in the wall of the cave. It continued to curve around. There were no antlers or horns. There was no long snout. There were no extended crests of bone. Just the rounded, beautiful simplicity of a hominid skull. The realization dawned that he had found it—the long-sought-after skull of Peking Man. But his elation quickly subsided as he began to contemplate the enormity of the responsibility now resting on his shoulders.

Pei found himself at the bottom of a long and rough tunnel with a priceless and exquisitely delicate fossil that could break into hundreds of unidentifiable shards if not handled exactly right. Night was falling and since Pei and the workers had been in the cave since early that morning, they were tired. He would have liked to cover the fossil hominid and come back in the morning when he was fresh, but it was too dangerous. A loose rock could fall on the skull or somebody might even slip in overnight and try to take it. He had to push on. Concentrating and lighting more candles, Pei worked on into the night, removing the skull in two pieces, carefully gluing the fragments and leaving as much of the adhering cave sediment in place as he could for support. He applied plaster bandage supports and waited for them to set. Then the pieces were slowly passed hand to hand up out of the cave. Pei took them to the field building and immediately set them close to the fire so that the glue and plaster would dry and harden.

The next morning Pei hurried to the train depot at Zhoukoudian to send a telegram to Davidson Black and to dispatch letters to Dr. Yang in Beijing and Dr. Weng of the Geological Society that a hominid skull had been found at Longgushan. Back in the office he wrapped the two fossil/sediment pieces in Chinese cotton paper and then covered them with burlap soaked in flour paste to support them on the outside. It was so cold that even in the relatively warm office, the burlap casings would not dry. Finally, on the third day, Pei put three heaters next to the skull pieces and they hardened.

When he was ready to go he put cotton padding around the fossil and then covered the whole with a quilt, disguising it as regular baggage. He hoped that by hiding the priceless fossil in this way, he could pass unnoticed at the various checkpoints on the road to Beijing. He started from Zhoukoudian by train early on the morning of December 6 and arrived at Beijing, some 40 kilometers away, a little before noon. He went straight to Davidson Black's laboratory to deliver the fossil skull.

Black was elated when he laid eyes on the fossil that Pei placed before him, in perfect, if as yet unprepared, condition. This young Chinese colleague of his was now beaming with delight and he was finally able to breathe again after handing over his treasure. Pei had just delivered Black from scientific limbo and had ensured Black's apotheosis in the firmament of paleoanthropology. And Davidson Black knew it. He was unstinting in his praise, realizing only too well what skill and fortitude it had taken to make this discovery. He made sure that the Chinese Geological Society, which later decided to award him a medal for the discovery, also struck one for Pei. And Black arranged for the Survey to publish Pei's own account of the discovery in its *Bulletin*.[33]

Black's trained anatomical eye hungrily scanned the archaic curve of the low skullcap, and the primitively jutting prow of the browridges, even as his political mind excitedly began to compose the letters he would write to the Rockefeller Foundation and his colleagues abroad. His risky but calculated naming of *Sinanthropus pekinensis* had yielded him two years of funding from the foundation, and now that risk had paid off. He could barely wait to get his hands on the specimen. It was beautifully primitive.

After Pei had carefully finished his cleaning and hardening of the specimen, Black went to work. He isolated each bone, ensured that the broken edges were free of all adhering matrix, and carefully rearticulated each bone into a composite whole. He worked for three months, making cast copies of the skull at each stage of the reconstruction. Black's original casts, signed in plaster by him on the back, are still in Beijing, stored now at the Institute of Vertebrate Paleontology and Paleoanthropology. Even before he was finished, he produced three preliminary papers on the new skull in 1930. His main paper on Peking Man's skull was to appear the following year.[34] But, by this time, a second, more fragmentary skull of the Longgushan hominid had been found in the 1930 excavations, and Black also included this specimen in his report.

The recovery in 1929 of what was to become "Skull III," in the terminology of the Peking Man fossils, marks a major turning point in the history of the Zhoukoudian site, as well as in Davidson Black's career. Discoveries of major significance now began cascading rapidly. In 1930 a number of teeth were found as well as another skull. In 1931 major discoveries of stone tools and evidences of fire at Zhoukoudian were made. In 1932 a well-preserved jaw bone of *Sinanthropus* was discovered. This specimen was to be the last fossil hominid from Zhoukoudian that Davidson Black was to study.

Limited only by how fast he could work and how much he could organize, Black had achieved everything he could have dreamed of. After a whirlwind tour of the Middle East and India, a return to his native Canada, and a trip to London to address the Royal Society, into which he had just been inducted, he returned to China in the autumn of 1933. Exhausted, he still went to Zhoukoudian at the end of the 1933 excavations. At the cave he collapsed but then continued with his examination of the site. When he returned to Beijing, he secretly went to the hospital, where doctors confirmed that he had suffered a mild heart attack. He kept his condition from even his wife. But in February he was hospitalized in Beijing for three weeks. Aware that his father had died of a heart attack at the age of 49 (Black was then four months shy of his forty-ninth birthday) and that his prognosis was assessed as "grave," he seems to have decided to die at his workbench. At about 5:00 P.M. on March 15, 1934, Davidson Black went into his laboratory, intending to work all night, as was his habit, for the first time since he had been released from the hospital. He reportedly chatted cheerfully with colleagues in the department before going to work at his desk. One of his last visitors was Dr. Yang, who recounted that he "found him sitting at his desk where he had worked for years and years at science. He talked of his anxiety as to whether his plans for the Cenozoic Research Laboratory could be carried out."[35] These anxious thoughts were close to Davidson Black's last. When Associate Professor of Anatomy Paul Stephenson came in around half an hour later, Black, still dressed in his white lab coat, was slumped over near his desk. He had died dramatically flanked by two of his greatest discoveries—*Sinanthropus* Skull III and the skull of *Homo sapiens* from the Upper Cave at Zhoukoudian.[36] His last paper, his lecture to the Royal Society in London, was to be published later that year, and it would be his last word on his fossils.

Amid all the furor and thrill of discovery of Davidson Black's last years, there had dawned, as well, a profound realization that Longgushan was beginning to reveal a very different version of human evolution than what Black or any of his colleagues had expected. As wonderful as *Sinanthropus* was, the species did not have the capacious brain box and aquiline features that Black and his mentors back in England had expected in this ancient ancestor of humanity. *Sinanthropus* was a far cry from Piltdown. But long-held preconceptions die hard. As the discoveries were continuing to come out of Dragon Bone Hill, someone needed to take up the torch for the fallen Black, describe the anatomy of the new fossils, and make some sense of it all.

A "First Class Man" to Carry on the Work at Zhoukoudian

The loss of Davidson Black, the charismatic leader of the Zhoukoudian research effort, could have spelled the end of the excavations. But such was the loyalty of those with whom he had worked and such was the productivity of the Zhoukoudian site that work was continued. The Rockefeller Foundation, for which hominid evolution has never been a major focus, continued to fund the excavations, probably out of loyalty to Black and his integration of the research with the medical school. And just as importantly, the foundation funded the position of anatomist to study and describe the fossil hominids that were still being discovered. But the search for a scientist who could fill the shoes of Davidson Black would be difficult. His was indeed a hard act to follow.

Immediately after Black's death his long-time friend and colleague, Pierre Teilhard de Chardin, temporarily took over the work at Zhoukoudian. In a letter to Walter Granger of the American Museum of Natural History on March 19, 1934, four days after Black's death, he wrote, "I have lost more than a brother. And the scientific work, in China, is deprived of half its soul."[87] Teilhard worried about where "an anthropologist of Black's standard" might be found to replace him. It must be a "first class man," and he asked Granger for suggestions. Teilhard started the Zhoukoudian excavations with Pei the next month.

Germany was one of the most active seats of physical anthropological and anatomical research in the early part of the twentieth century. In the 1930s the country was also in economic and political turmoil, with many of its most prominent professionals fleeing the antiintellectual and ethnic persecutions of the National Socialists. These two factors—German eminence in the anatomical sciences and mass emigration to escape Nazi control—united to supply one of the most prominent German researchers, Professor Franz Weidenreich of the University of Frankfurt-am-Main, to be Black's replacement. William King Gregory, curator of anthropology at the American Museum of Natural History and Granger's associate, was mentioned in Teilhard's letter of entreaty from Beijing. This accomplished anatomist and sometime protégé of museum director Osborn may have played a part in bringing Weidenreich and Zhoukoudian together. A few blocks away, in midtown Manhattan, the China Medical Board of the Rockefeller Foundation was forming a committee to replace Davidson Black.

In April 1933 Germany's National Socialist government of Adolf Hitler, assuming broad police powers, dismissed all Jews from university posts. Weidenreich, who was a full professor of anatomy and who was also

ethnically Jewish,[38] found himself suddenly dispossessed of his professorship and his country. One can only imagine his bitterness. He was 60 years
old and had contributed a full career of academic, medical, and political
leadership to Germany. But Germany's loss was world science's gain. In
1934 he left, never to return, to accept a visiting professorship at the University of Chicago. His exposure to American colleagues brought him to
the attention of the Rockefeller Foundation. In 1935 the foundation named
him visiting professor of anatomy at its medical school in Beijing and
honorary director of the Cenozoic Research Laboratory, the posts left va-
cant by Davidson Black.

Davidson Black never met Franz Weidenreich, a contemporary nine
years older than he. The closest they came to meeting was in 1914, on the
eve of World War I, when Black was studying neurology and fossil brain
endocasts with Dr. Ariens Kappers in Amsterdam, and Franz Weidenreich
was in Alsace-Lorraine as professor of anatomy at the University of Strassburg.
Weidenreich had received his M.D. from Strassburg in 1899, writing his
dissertation on the cerebellum of living mammals. He then advanced through
the academic ranks under the tutelage of his professor, the legendary Gustav
Schwalbe.[39] When Schwalbe retired in 1904 the young Weidenreich was
named to replace him as professor of anatomy at Strassburg. He spent the
next ten years building a solid body of research on blood cells, skeletal
tissue, overall skeletal form, and human evolution, studying and describ-

Franz Weidenreich at Dragon Bone Hill on November 16, 1936. This photograph was
taken during a visit to the excavation in November 1936, after the recovery of Skull X.

ing a number of European fossil hominids. But the onset of World War I caused upheaval in Weidenreich's professional and personal lives, and he became active in politics. An ardent German, he dropped scientific work for a number of years, became president of the Democratic party of Alsace-Lorraine, and served as a member of the Municipal Council of Strassburg from 1914 to 1918. When France was victorious in 1918 and took over Alsace-Lorraine, Weidenreich was relieved of his university post and he and his family fled into Germany. It took three years for Weidenreich to regain an academic position, this time at the University of Heidelberg. Heidelberg was the home of a famous hominid fossil, "the Heidelberg Jaw," found in 1907 in a gravel deposit at Mauer and for many years Europe's geologically oldest human fossil. A study that he published in 1926 on a Neandertal fossil skull from Weimar-Ehringsdorf, near Goethe's former haunts, brought him recognition in Frankfurt, the city most associated with Germany's great poet and naturalist. Weidenreich was offered the professorship of anatomy at Frankfurt and moved there in 1928. It was while he was at Frankfurt that Weidenreich first read of Davidson Black's discoveries in China. He quickly appreciated the similarities between the Zhoukoudian discoveries and such specimens as *Pithecanthropus* from Java and the Mauer mandible.[40]

Franz Weidenreich was described by William King Gregory as "the flower of German civilization and true culture." But in 1934 he made a total break with his homeland, refusing even to publish in German. After 1935 every one of his 48 papers and books was in English, whereas 143 of his 144 publications before 1935 were written in German. Weidenreich was able to get his wife, Mathilde, and one daughter out of Germany and to China with him, but his two other daughters and Mathilde's mother were sent to concentration camps. While he focused on the ancient hominids from Zhoukoudian in his work, a pall hung over Weidenreich's personal life. He worked for years to gain release of his family from Germany, and he eventually did succeed in reuniting with his daughters years later in the United States. Tragically, his mother-in-law died at the hands of the Nazis, and one of his sons-in-law was shot.[41]

Franz Weidenreich arrived in Beijing in April 1935, thirteen months after Davidson Black had succumbed at his workbench. The 1935 excavation season had already begun. Pei and excavation chief Lanpo Jia were competently running what had become a well-tuned machine, discovering new fossils of all types of mammals, including hominids, at a good clip.[42] Teilhard and his paleontological colleagues worldwide were in the wings, working to ensure that whatever was found at the cave would be immediately interpreted in light of the most current paleontological knowledge. Black's support staff at Peking Union Medical School were all still in place. Weidenreich just had to walk in and assume Black's role of

View of the excavation in the spring of 1935, looking toward the southeast. The village of Zhoukoudian is in the background. The roof of Pigeon Hall Cave is just visible in the upper left corner under the plank walkway. The number "58" in the center of the photograph indicates that this was serial field day 58 of the 1935 field season. Records show that excavators were working in Level 11 of Layer 8/9. The grid system of one-meter squares can be seen painted on the walls. Blocks of 4 square meters each were excavated at a time.

paleoanthropologist and interpreter of the hominids. At this he was to prove masterful.

The 1935 field season at Zhoukoudian was very productive. More of *Sinanthropus* Skull V had been discovered by Jia.[43] Teilhard, who had been posted back to France for three months, relayed from Paris on July 25 that "Weidenreich is acting in a wonderful way: quiet and positive. Yet, we miss terribly Davy [Davidson Black]."[44]

Excavations, again funded by the Rockefeller Foundation, were resumed at Zhoukoudian in spring 1936. The digging was halted during the heat of the summer but resumed in September, this time for the last time. Three new hominid partial skulls (Skulls X, XI and XII) were discovered along with isolated teeth, again by Jia.[45] Guerrilla fighting broke out in the Western Hills, where Zhoukoudian was located, and the work at Zhoukoudian had to be abandoned.

Much of Weidenreich's job in China was not the discovery of new fossils at Zhoukoudian, but the study, detailed anatomical description, and interpretation of all the riches that had been found since the 1932 jaw, the

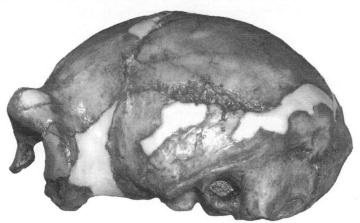

Top: Chief Excavator Lanpo Jia in Locality 1 cleaning the third of the *Homo erectus* skulls (Skull XII) that his team discovered in November 1936. This area (Square I, 2) and stratum (Layer 8/9, Level 25) is part of Locus L, which yielded a total of four hominid individuals. *Bottom:* A side view of Skull XII, likely an adult male (photograph of a first-generation cast).

last specimen described by Black. Teilhard, the tireless correspondent, wrote from Beijing in early 1936 that "Weidenreich is studying *perfectly* the old and new material of *Sinanthropus*, and reaches many new, well based, conclusions concerning the exceptionally primitive characters of the form."[46] But time was running out.

The demise of the Cenozoic Research Laboratory, one of Black's last anguished worries, finally occurred in December 1941 with the capture of Beijing by the Japanese army.[47] Franz and Mathilde Weidenreich had left Beijing in April, taking with them plaster casts of the Zhoukoudian specimens and Weidenreich's copious anatomical notes on the original specimens. They went to New York where Weidenreich was given a visiting (unsalaried) appointment at the American Museum of Natural History through Gregory's enthusiastic intervention. Henry Fairfield Osborn had died in 1936, but he would have been gleeful that his museum eventually received the describer of the hominid fossils from Zhoukoudian, the cave site from which J. Gunnar Andersson had so adroitly outmaneuvered the American Museum team years before.

It was in New York between 1941 and 1948 that Weidenreich completed his series of monographs on the Zhoukoudian hominids, securing for them a place in human evolutionary interpretation. Weidenreich became the interpreter of one of the most compelling epics in human evolution.

By the middle of the twentieth century, Peking Man was a household phrase worldwide. But though Weidenreich did a masterful job of spreading the fame and enshrining the memory of Peking Man, he was unable to ensure the curation of the physical remains of the actual hominid fossils from Longgushan Cave. When the director of the Chinese Geological Survey wrote in 1941 from the provisional capital of Chungking (Chongqing) to ask Weidenreich, who was still in Japanese-occupied Beijing, to take the fossils with him to New York, Weidenreich was forced to decline. The president of Peking Union Medical College, Henry Houghton, had decided not only that it was time for Dr. Weidenreich to leave China[48] but that he was not to take the Peking fossils with him.[49] When Weidenreich departed Beijing, leaving the priceless fossils in their storeroom at Peking Union Medical College, it would be the last he ever saw of them. Houghton had made a fateful decision.

CHAPTER 2

The Dragon Reclaims Its Own

On July 7, 1937, at the Marco Polo Bridge on the road between Beijing and Zhoukoudian, the Imperial Japanese Army fired on Chinese civilians in an incident that exploded into the Sino-Japanese War. Excavation at Longgushan ceased two days later as the turmoil spread across northern China.[1] Head excavator Lanpo Jia directed the workers to disperse and seek safe haven in Beijing or elsewhere. Most did, but 26 workers who lived in the town of Zhoukoudian stayed on at the site to keep an eye on the excavation, the buildings, and the equipment. They were still on the payroll of the project at the end of 1937.[2]

The Japanese army soon conquered all of the area of northern China around Beijing, including Zhoukoudian. But the small numbers of troops left to control a restive population were insufficient to ensure calm. Communist guerrilla militias sprang up around the country to fight for an independent China. One such group became ensconced at Zhoukoudian, virtually under the noses of the Japanese High Command in Beijing.[3] Many locals rallied to surreptitiously support the guerrillas' cause, and three excavation workers at Longgushan—Wanhua Zhao, Zhongyuan Dong, and Yuanchang Xiao—worked in the kitchen for some one hundred soldiers occupying the old temple building and other buildings at the site. A number of skirmishes between the guerrillas and the Japanese army occurred throughout 1937 and early 1938, but by April Japanese plainclothes troops had occupied Longgushan. Soon thereafter, on May 3, 1938, Zhao, Dong, and Xiao were arrested by the Japanese and taken to their headquarters at Fangshan. There they were interrogated and tortured in attempts to extract information about the guerrillas' movements and whereabouts. News of their deaths reached Lanpo Jia in Beijing by a messenger from Zhoukoudian. They had

been bayoneted to death along with some 30 other prisoners. Contemporary Japanese accounts confirm that recruits in the Japanese army serving in China were routinely required to bayonet prisoners in order to harden them for battle.[4] Lanpo Jia recorded Teilhard's reaction to hearing the news: "He immediately stopped typing; his face turned pale, his lips trembled, and his eyes stared at me. He sat motionless for a while, then slowly stood up, and with his head bending low, began to pray."[5] In 1946 the International War Crimes Tribunal convicted a number of Japanese generals and lower-ranking officers of war crimes and sentenced them to death. It has been estimated that some ten million Chinese civilians were killed by Japanese forces between 1936 and 1945.[6] This horrendous loss of human life was the tragic backdrop for what happened to the long-dead fossils of Peking Man.

Peking Man Under Siege

Soon after the incident at Marco Polo Bridge and the cessation of excavation at Longgushan, in July 1937, Franz Weidenreich asked his technical assistant at Peking Union Medical School, Chengzhi Hu, to begin packing up all the hominid fossils. Hu enlisted a carpenter to make two crates for the fossils. He then carefully wrapped each fossil in layers of protective paper and cotton batting, made a packing list, and placed them into the crates. Once the crates were packed, Weidenreich had them delivered to the vault of an American bank in Beijing for safekeeping in case the Japanese took over the medical school.[7] How long the fossils stayed in the bank vault is difficult to determine, but at some point they were returned to Peking Union Medical College, perhaps in the latter part of 1937 when it became clear that the Japanese occupiers of Beijing intended to respect the territorial concessions of the various foreign governments in China. But Hu kept the crates. They were to be used again.

That there were others interested in the Peking Man fossils and their site was evidenced by an incident that occurred in late 1937. The head

---

Facing page
View of the excavation at Locality 1, looking east, June 15, 1937. The vertical opening of Pigeon Hall Cave can be seen at the upper left. This photograph was taken on serial field day 165, approximately a month before excavation was halted by the onset of the Sino-Japanese War and subsequently by World War II. The stratum excavated here was Level 28, Layer 10. In the upper left of the excavated area (Square K, −2), two fossil teeth of a macaque monkey were discovered. Immediately below this level (Square H, −4, Level 29) the last *Homo erectus* skull (Skull XIII) was discovered.

technician at Longgushan, Wanhua Zhao, reported in a letter written on November 10, 1937, to Lanpo Jia back in Beijing that three truckloads of Japanese soldiers and six civilians had arrived at the site. The civilians (who we suspect were Japanese academics) had come armed with technical articles about Dragon Bone Hill geology and paleoanthropology; the soldiers had come armed with rifles. They asked the whereabouts of Pei and Lanpo Jia. They took photos and some measurements of the site, had a picnic lunch, and then left in the afternoon. In his book, Lanpo Jia cites this invasion of a well-established site as proof that the Japanese were after the Peking Man fossils. But what if they had been invited?

A report by Warren Weaver of the Rockefeller Foundation dated June 20, 1941, provides the details of an amazing interview with Franz Weidenreich soon after the latter arrived in New York. It places the incident at Longgushan in a different light.[8] The report states that Weidenreich "has been on good terms with an elderly Japanese archaeologist, Professor R. Torii,[9] who is now staying with his family at Yenching University."[10] Even though Weidenreich was no longer a German citizen,[11] he may have been given the latitude and special considerations accorded the German community in China by the Japanese, who now were Axis allies of Germany. Weidenreich likely had relatively free access to other academics in Beijing during the Japanese occupation. Weaver's 1941 report says that Weidenreich "thinks that it would be entirely feasible to carry on the Cenozoic program [at Zhoukoudian] actively at the present time." Could Weidenreich have been aware of the Japanese visit to Zhoukoudian in 1937? Was it he who supplied the publications to the Japanese visitors? Although we have no evidence that Torii was one of the civilian visitors to the site, Weidenreich may well have contemplated and even begun setting up a collaboration with the Japanese to excavate more fossils, but he may have diplomatically kept this information from his Chinese colleagues. Certainly, Lanpo Jia never makes any mention of Professor Torii or Weidenreich's possible connection with him.

We know of Weidenreich's opinions regarding the Japanese from Dr. Weaver's notes from the 1941 meeting. Weidenreich strove to convince the Rockefeller Foundation of the feasibility of reinitiating excavation at Dragon Bone Hill under Japanese occupation. Indeed, this was one of his primary objectives in meeting with Dr. Weaver in June of that year. Weaver writes that "W[eidenreich] thinks that it would be entirely feasible to carry on the Cenozoic program actively at the present time. It would, indeed, be necessary to ask permission of the Japanese authorities. But W[eidenreich] feels confident that this would be granted and that no difficulties would arise." Dr. Houghton opposed the plan and had successfully blocked it. Weidenreich complained to Weaver that Houghton was determined to make a political issue out of continued research at the cave. Weaver quotes

Weidenreich as saying that "it would be perfectly simple to keep the work quite innocently removed from politics." Houghton, on the other hand, reportedly countered that the Chinese members of his board would resign in protest if he were to ask permission from the Japanese or appear to collaborate with them in any way. Weidenreich's last word with Weaver was that "this is entirely unrealistic since the Japanese are actually in control and their permission must be obtained on various matters."

Whatever Weidenreich's intentions to reinitiate research at Longgushan during the Japanese occupation of Beijing before the Pearl Harbor attack, between 1937 and 1941, they bore no fruit. In retrospect, however, Weidenreich's opinion that the Japanese posed no real threat to the safety of the Peking Man fossils helps to explain why, after the fossils returned to Peking Union Medical College from the Beijing bank vault, no action was taken to safeguard them for the four long years of Japanese occupation. They went back into the safe in Weidenreich's office, and he continued to examine, measure, compare, and describe the fossils. Technicians worked feverishly to finish the molding and casting of the fossils so that accurate replicas could be made and sent abroad. Artists and photographers worked closely with Weidenreich to render the fossils carefully for his illustrated monographs. In the spring of 1939, Ralph von Koenigswald, a German paleontologist who had discovered new *Homo erectus* fossils in Java, visited occupied Beijing and brought his fossils with him. He worked with Weidenreich in the medical school for two months, comparing and contrasting the early hominid specimens from the two most important fossiliferous areas of Asia. This period of comparative study was to prove seminal in the interpretation of the evolutionary significance of *Homo erectus* by both scientists, but privately von Koenigswald was criticized by Weidenreich's staff. Hu is quoted as saying, "We all worried about the safety of the specimens he brought along and disapproved of his incautiousness."[12] Yet at the war's end all of von Koenigswald's fossils were accounted for. A much different fate awaited Peking Man.

The Nationalist Chinese government under Chiang Kai-shek had moved south to the city of Nanjing

German paleoanthropologist Ralph von Koenigswald (*center*) in 1938 with Pierre Teilhard de Chardin (*left*), visiting from Beijing, and German geologist Helmut de Terra (*right*) in Java.

(then "Nanking") in July 1937 when the Japanese took Beijing. The Chinese Geological Society, which was the agency of the Chinese government responsible for the Peking Man fossils, moved with some of its collections and most of its personnel to Nanjing as well. It left the Peking Man fossils behind, in the belief that they would be safe under the aegis of the American-backed Peking Union Medical College. Nanjing fell to advancing Japanese forces in December 1937, in a rout known variously as the "Massacre of Nanking" or the "Rape of Nanking." Some three hundred thousand Chinese civilians were killed, and the members of the Nationalist government fled to the western city of Chungking. The Peking Man fossils stayed securely in their safe in the medical school in Beijing during this tumultuous time.

The director of the Geological Survey of China, Dr. Wen-hao Weng, despite the disorganization of his activities caused by the war, continued to take his duties regarding the safety of the Peking Man fossils seriously. It was he, along with Davidson Black and Zhongjian Yang, who an excited Wenzhong Pei cabled when he had discovered the first skull of Peking Man back in 1929. Weng, based now in Chungking, wrote a letter to Dr. Henry Houghton, president of Peking Union Medical College, on January 10, 1941.[13] In it he says that "we are ready to agree to have them [the fossils] shipped to America and entrusted to some scientific institution for temporary safe-keeping during the war period in China after which they should be returned." The letter arrived while Houghton was in Shanghai and he did not read it until his return to Beijing. Weidenreich had already been advocating such a course of action. On April 10, 1941, Houghton writes,[14] "Some weeks ago Dr. Weidenreich raised the question with me as to whether or not it might be possible or practicable, with the consent of the officials of National Geological Society and of the Chinese National Government, to remove the human material and artefacts to some one of the great museums in the United States, there to be held in custody for the duration of the war." After talking to the U.S. embassy in Beijing, he says rather brusquely that he "came to the conclusion that it would not be in order to do so." Weng's letter then reopened the issue.

Lanpo Jia and the Chinese researchers who had worked at Longgushan naturally considered Franz Weidenreich, professor of anatomy and the inheritor of Davidson Black's honorary directorship of the Cenozoic Research Laboratory, to be the man in charge of deciding whether the Peking Man fossils should stay in China or go to the United States for safekeeping. They had no way of knowing that an internal power struggle was going on between Weidenreich and Henry Houghton. Although Weng had also written to Weidenreich and Pei about the need to move the fossils either to Chungking or to the United States, it was Houghton who had to make the final decision. Weng wrote to Houghton that he had asked Weidenreich and Pei "to consult your [Houghton's] opinion for an early decision and mak[e] all necessary arrangements on our behalf."[15]

Henry S. Houghton had come to Beijing in 1918 from Johns Hopkins University School of Medicine in Baltimore to head Peking Union Medical College. It had been he who years previously had attempted to dissuade Davidson Black from researching fossils and "mysterious caves." In his reply to the Rockefeller Foundation written on April 10, 1941,[16] it is clear that his low opinion of the value of the Peking Man fossils had not changed. Houghton, in discussing the question of safeguarding the most important collection of fossil hominids at that time in existence, describes the fossils as "somewhat parallel . . . to our unique collection of Chinese medical books." In turning down Weng's request to move the fossils out of Beijing, Houghton gave a number of reasons: (1) the Japanese had control over northern China and would not recognize any agreements or permissions from the Nationalist government, (2) the Japanese controlled all customs inspections of shipments leaving Beijing, (3) the fossils would likely be seized if an attempt were made to export them secretly, and (4) the U.S. government, for whom Houghton presumed to speak on the basis of his contacts with the Beijing embassy, "cannot in the nature of the circumstances extend any aid or countenance to the removal of property to which the Chinese National Government has title." Houghton's rationalizations for doing nothing about the Peking Man fossils were based on his real reason, his fifth point, quoted here in its entirety because, as history was to show, it was so utterly incorrect:

> On the other hand, it does not seem to me that these specimens, unique and valuable as they are, are in particular danger of destruction if they remain in the custody of the College. They have no sale value and at the worst could only be confiscated and taken elsewhere. In such cases the end result would be negotiations for their return to the Chinese government to which they belong, and a judgment on such a matter must necessarily be held in abeyance until we know more about the end results of current hostilities.[17]

Houghton replied to Weng that removal of the Peking Man fossils was "wholly out of the question."

Franz Weidenreich had, since early 1941, written a series of letters and memoranda to Houghton arguing for action on the Peking Man fossil problem. By April, Houghton had had enough. He dispatched Weidenreich back to New York and sent his letter of April 10, 1941, to the chairman of the China Medical Board, with him.[18] Weidenreich took his research notes and plaster casts but left his library and the original fossils in the medical school. He clearly intended to come back, but he was not to see China again.

In public, Weidenreich said to all his medical school and research colleagues that now that he had completed his primary observations on the fossils, he would go to the United States to complete his monographs on Peking Man. He personally took responsibility for the decision to leave the Peking Man fossils in the medical school, even though he had fought in private with Henry Houghton and with the U.S. embassy for a positive response to Dr. Weng's request. His public statement that he could not ensure the safety of the fossils since he did not travel in any official governmental capacity was true. Even his citizenship was in doubt.[19] The embassy had apparently agreed to consider him a U.S. citizen, but there was a question of whether Houghton would support him.[20] Weidenreich was perhaps also mindful of the criticism voiced against von Koenigswald when he brought the Javanese fossils to Beijing. The Rockefeller Foundation agreed to continue Weidenreich's salary and he was offered a visiting position at the American Museum of Natural History in New York. A large send-off party was held in his honor at Lockhart Hall of the medical school and soon thereafter he sailed for the United States. By early June he was in New York. Wenzhong Pei took over the administrative responsibilities for the Cenozoic Research Laboratory.

In July 1941, soon after Weidenreich had arrived in New York, U.S. intelligence broke the Japanese diplomatic code. American officials learned from intercepted correspondence that the Japanese were planning a major escalation of the war in China and that they intended to advance southward into Indochina and Thailand later in the year. Any pretense of avoiding confrontation in China was now lost. Perhaps partly for this reason, partly because Dr. Weng of the Geological Survey made a direct plea to the U.S. ambassador, and partly because the president of the Rockefeller Foundation, Raymond Fosdick, had agreed to "talk over with his friends in the State Department the possibility of safe removal of the Cenozoic material,"[21] the Americans decided in September 1941 to provide safe transport for the Peking Man fossils out to a temporary home in the United States. Apparently, it was not a priority, however, particularly since Dr. Houghton was locally in charge of the operation; nothing happened for some three months. Then in late November 1941, Ms. Claire Hirschberg (later Taschdjian), Weidenreich's secretary, told Mr. Hu that he should box up the Peking Man fossils for shipment. Mr. Hu confirmed this with Dr. Pei and began the work with an anatomy department technician, Mr. Yanqing Ji.[22]

Where the Peking Man Fossils Went

The last anyone reliably laid eyes on the original Peking Man fossils was during the packing of the specimens by technicians Hu and Ji, reported by journalists Ming-sheng Li and Nan Yue in a large compendium of the status of the search for the missing fossils published in Chinese in 2000.[23] Hu recounted in a letter to Lanpo Jia in 1977 how the fossils had been packed: "We wrapped every fossil in white tissue paper, cushioned it with cotton and gauze and then over-wrapped them with white sheet paper. The packages were placed in a small wooden box with several layers of corrugated board on all sides for further protection. These boxes were then put into two big unpainted wooden crates, one the size of an office desk, the other slightly smaller."[24] He then added, "We delivered the two cases to the head of Controller T. Bowen's office, at the Peking Union Medical College, and from then on none of the Chinese knew what happened to them." Controller Trevor Bowen's office was in Building B of the medical school.

There was a reason that the Chinese members of the research project were kept in the dark. As Pei later wrote, "We should be grateful to our American friends, who not only had assumed the entire responsibility for transporting 'Peking Man,' but also were prepared to shoulder the blame if they should become prisoners after the war broke out between Japan and

The front gate of Peking Union Medical College, Beijing, in the late 1930s. It was through this gate that Medical College Controller Trevor Bowen wheeled the crated Peking Man fossils to a waiting car around November 20, 1941. The fossils were reputedly delivered to the United States Legation for delivery to the U.S. Marine Corps and shipment to the United States, but no reliable accounts exist to show that they were ever seen again.

the United States so as not to implicate Chinese (namely myself)."[25] He mentions Mr. Trevor Bowen, the controller of the Medical School, and Dr. Henry Houghton, as the primary individuals who "took care of the packing . . . as well as their transportation." The fact that there was an intermediate step between Hu and Ji's packing the fossils and the fossils leaving the Peking Union Medical School is an important detail.

According to Jia and Huang's 1990 book, Wenzhong Pei recollected that the crated fossils were moved to a strong room in another building at Peking Union Medical College—Strong Room 4 in the basement of Building F—between November 18 and 20, which was 18 to 20 days before the Japanese attack on Pearl Harbor. Professor Wenzhao Ma and a worker were identified as the individuals who transported the two boxes by wheelbarrow. Jia and Huang then state simply that "it is known that on the day following the packing, the fossils were delivered to the U.S. embassy located at Dongjiaominxiang in Beijing, and since then they have been missing." No attribution is given for this information. Dr. Harry Shapiro reported in 1974 that a secretary at the medical school, Miss Mary Ferguson, had written to him to say that she had seen Mr. Bowen, the controller, taking a trunk across the marble courtyard to a waiting car at the front gate. She stated that it then "went to the U.S. Marine barracks,"[26] not to the U.S. "embassy" (actually a legation, a lower-level diplomatic presence), but it is unclear how she knew this. The U.S. legation was next door to the U.S. marine barracks in Beijing.

A very different story is given in Li and Yue's 2000 book.[27] According to Li and Yue, the fossils stayed in Strong Room 4 in Building F for approximately two weeks, but during that time they were repacked into redwood boxes, apparently by Mr. Bowen. No Chinese eyewitnesses attest to this repacking and no independent records from the medical school or the Rockefeller Foundation have been located to confirm this. However, this book reports a recently discovered 1945 interview of two U.S. Marine prisoners of war in Japan—a Sergeant Snider and Sergeant Jackson—who claim to have picked up two redwood boxes at Peking Union Medical College on orders from their commanding officer, Marine Lieutenant MacLiedy. Snider and Jackson said that they believed the boxes contained the bones of Peking Man. According to their detailed recounting they picked up two boxes on December 4, 1941, by truck and delivered them to the lieutenant at the Beijing U.S. marine barracks that day. They were then ordered to take the boxes to the Beijing train station the next morning and guard them all the way to their destination—the Swiss warehouse at the port city of Qinhuangdao where they were to await transport by ship to the United States. The marines reportedly arrived at Qinhuangdao late on the afternoon of December 5, deposited the boxes safely, and left by rick-shaw to spend the night at nearby U.S. marine Camp Holcomb. They returned to Beijing by train the next day.

There are still other versions of what happened to the fossils. In one story, discussed in Shapiro's 1974 book and promulgated by former Marine Captain William Foley, M.D., a neighbor of Pierre Teilhard de Chardin in Beijing and later a New York cardiologist, the fossils were packed in his personal baggage for transport to the United States. He claimed that the baggage with the fossils had been sent by train to Camp Holcomb, where he and his detachment were all captured on the morning of the attack on Pearl Harbor (December 8, 1941, Beijing time). Other reports claimed that the fossils never made it to Camp Holcomb or to Qinhuangdao, but had been captured by the Japanese en route and thrown out by troops ransacking the train, who were unaware of their significance. Other reports that the fossils had been loaded onto the scheduled ship, the SS *President Harrison*, which had then been sunk, were found to be erroneous. Records showed that well before the *Harrison* ever reached port it had been intentionally run aground at the mouth of the Yangtsze River by the captain to avoid capture by the Japanese.[28]

Deciding which of these conflicting versions of the disposition of the Peking Man fossils is more likely is difficult, but discrepancies in some of the stories make some less plausible than others.

William Foley's claim that the fossils were packed in glass jars makes his story unlikely. This manner of packing would be very unusual for paleontological specimens and does not match Mr. Hu's detailed description of how he had wrapped and packed the fossils. Glass jars seem to have been first mentioned by Colonel William Ashurst, Dr. Foley's superior and commander of the U.S. Marine detachment at the Beijing U.S. embassy, to whom the fossils had reputedly been entrusted. Author Ruth Moore in her 1953 book *Men, Time, and Fossils* states that Dr. Henry Houghton made the request of Colonel Ashurst. Dr. Foley had demanded from Lanpo Jia a meeting with top officials in China before he would discuss more details of what he might have known. Such a request struck Jia as arrogant and he was quoted as saying that it "made me really angry."[29] It also suggests an ulterior motive for Foley's claims regarding the high-visibility Peking Man fossils. In any event, Jia was unable to arrange such a meeting and Foley died in 1992 without divulging what if anything he might have known.

The story of the two marines interviewed in 1945 is not verified by another source. Jackson died of pneumonia in Japan and Snider died in an automobile accident in the United States after his release from Japan at the end of the war. Their story does accord with many other details, except

Chief Excavator Lanpo Jia at Longgushan on November 2, 1936.

that their description of the boxes as being made of redwood does not agree with Hu's reliable description of white unpainted boxes. And the marines never saw what the boxes that they had transported contained. The boxes picked up by these two marines could have been two unrelated shipments from the medical school to the United States.

It is worth noting the close similarities between the story related by the two marines and the supposedly fictional account written by Weidenreich's former secretary, Claire Taschdjian in 1977 (*The Peking Man Is Missing*). Taschdjian was in Beijing throughout the war years. Jia and Dr. Yang met her by accident on a Beijing street in 1947, when she told them that Japanese military police had arrested her and taken her to search a number of warehouses in Tianjin during the war.[30] In her "novel," two marines, one of whom is named "Snyder," conspire to switch the fossils with a shipment of laboratory glassware and hijack them to the United States. "Kathy," Taschdjian's semiautobiographical heroine, is responsible for inadvertently introducing the hijackers to the fossils in the medical school through her romantic involvement with one of the marines—the one who later died of pneumonia, both in the book and for real in a Japanese prisoner-of-war camp in Hokkaido. The fossils in Taschdjian's book were kept throughout the war in China by the wife of a heroin dealer who had masterminded the heist and then died unexpectedly. She then married a diplomat and eventually brought the fossils to New York when he was transferred to the United Nations. The fanciful finale of the novel includes the murder of the heroine and the accidental death of the surviving hijacker. The Peking Man relics end up being smashed in a garbage compactor by the heroine's disgruntled and superstitious landlady, and finally taken off by a garbage truck to the New York City dump.

Taschdjian's book is an imaginative attempt to make up a fleshed-out, plausible explanation that fits most of the facts of the Peking Man case. Even Christopher Janus's then recently published encounter (in a 1975 book by Janus and Brashler) with the mysterious "Empire State Lady," who claimed to have inherited the fossils as a legacy from her late husband, finds an explanation. In real life Claire Taschdjian, who died in 1998, believed that the fossils were eventually discovered at Qinhuangdao by Japanese soldiers. She thought that the Japanese may have considered the bones to represent the ultimate ancestors of the Chinese, and they dumped the relics into the bay as an insult to the country and its people.[31]

One final category of possibilities is that the Peking Man fossils were buried somewhere to prevent their discovery. Ralph von Koenigswald had buried most of his fossils in Java and managed to conceal them effectively from the Japanese army. Li and Yue reported an interview with a Japanese army doctor who claimed that he took part in burying the Peking Man fossils.[32] These had been found by the Japanese and were buried before they evacuated Beijing. Strangely, he added that the preserved internal organs of Chinese political leader Sun Yat-sen (who had died at Peking Union Medical College Hospital in 1925) were buried along with the Peking Man fossils. The location was about two kilometers east of the medical school near an old tree. Chinese authorities took his story seriously enough to excavate down to a depth of more than a meter and a half a large area around an old tree located two kilometers from the medical school. Nothing was found.

Another scenario was recently suggested by Christopher Janus, the U.S. entrepreneur who undertook a search for the missing Peking Man fossils in the 1970s.[33] Some time after Janus published his book he was contacted by a hospitalized man in Texas. The man, a Mr. Innes, identified himself as a former marine in China and he said that he had wanted to pass on information about the Peking Man fossils before he died. He recounted to Mr. Janus that he had been on guard duty at the Beijing marine compound one night just before the Pearl Harbor attack. While he was having a cigarette break two marines came through the gate around midnight carrying two trunks. They left shortly thereafter without the trunks. Innes presumed that they had buried the trunks somewhere within the marine compound. He also guessed that the trunks contained the bones of Peking Man. Mr. Janus was not aware of any systematic attempt to search the site of the former U.S. marine compound in Beijing for the missing fossils.[34]

This story of possible U.S. marine hijacking of the Peking Man fossils finds some support in an opinion expressed by James Stewart-Gordon, an editor of *Reader's Digest Magazine* who undertook research on the disappearance of the Peking Man fossils.[35] Mr. Stewart-Gordon notes that the U.S. marines stationed in China just before World War II were a rowdy lot, notorious for using drugs, taking "cumshaw" (bribes), and having the highest venereal disease rate in the U.S. military. It is not unlikely in his opinion that one or more of the U.S. marines assigned to guard the fossils actually absconded with them or perhaps switched footlockers. The marines may then have been taken prisoner and died during the war, taking the knowledge of the fossils' whereabouts with them. The scenario again recalls the plot of Claire Taschdjian's novel, it could explain the reticence of Dr. Foley to disclose everything he knew about the fossils, and it is not incompatible with the stories of the captured U.S. marines interviewed in 1945 in Japan. If such an event occurred it is easy to see why the United States would not be forthcoming with such embarrassing information.

Why Were the Japanese Interested in Peking Man, and Did They Find the Fossils?

On December 8, 1941, Pearl Harbor Day, most staff and employees at Peking Union Medical College were paid their salaries and dismissed. The next day, the Japanese occupied and took over Peking Union Medical College, posting guards at the gates.[36] Shortly thereafter, according to an interview with Wenzhong Pei,[37] Dr. Kotondo Hasebe, an anthropologist at Tokyo Imperial University, accompanied by his assistant Mr. Fuyugi Takai, "hurried to find the Peking Man fossils." Pei claimed that Hasebe had come to China "long before the Pearl Harbor attack," a reference to Hasebe's likely being one of the civilians who went with three truckloads of soldiers to the Longgushan site in 1937. In 1941 at Peking Union Medical College he came with soldiers of the Japanese army, and according to Pei's report, "when they ordered the safe opened and saw that there was a copy of the skull-cap, they left without a word." A few days later Pei was interrogated by a captain in the Japanese army who confiscated his resident identification card, essentially confining him to the Beijing city limits. Pei professed ignorance of the fossils' movements and current whereabouts, citing the distance of his office from the medical school. The captain told him that Americans at the medical school were suspected of smuggling the fossils out of China. Dr. Houghton, the president, and Mr. Bowen, the controller, had both been arrested and were in Japanese custody. The captain told Pei that he could continue working as usual unless the army authorities decided to press the search for the fossils. He added the threat, "If they do, you can't get away by pleading ignorance."

Kotondo Hasebe played an important, and as yet unappreciated role, in the continuing Japanese efforts to locate the Peking Man fossils. According to Pei, when soldiers first occupied the medical school in Beijing, Japanese authorities were primarily aware of its importance in the field of medicine, paying "only incidental attention to the problem of the Peking Man fossils." However, when Hasebe sent a report to the Ministry of Education in Tokyo, his information was forwarded to the emperor.[38] It was widely believed that it was Emperor Hirohito himself who then ordered the North China Expeditionary Force of Japan to reinitiate the search for the missing fossils. In April or May 1942, five to six months after Pearl Harbor, Pei was summoned by Japanese authorities to the Hotel Beijing for further questioning. He gave the same answers as before and was released. But on returning home he was questioned again and then placed under house arrest for two weeks by a Japanese military detective. Dr. Hasebe then reappeared, this time in the company of several Japanese army officers. They took Pei to Zhoukoudian and visited the site.

Hasebe told Pei that they were contemplating resuming excavation at Dragon Bone Hill.

Dr. Kotondo Hasebe appears from his publications to have been primarily an ethnologist, specializing in Micronesia. He wrote papers on customs of the Marshall Islands in 1915[39] and on body ornamentation, especially tattooing, in various Micronesian cultures between 1917 and 1943. All of these papers were published in one journal, the *Journal of the Anthropological Society of Tokyo*. With his background, why Hasebe doggedly pursued Peking Man from 1937 to 1943 is mysterious. Perhaps it was a wartime duty assigned to him, or perhaps it was an abiding personal interest and he was able to gain the ears of influential persons in Tokyo to assist him. Perhaps he was acting on behalf of a powerful patron or patrons in Japan, close to the emperor, who had an interest in the Peking Man fossils. Whatever the source of his apparent interest in the fossils of Peking Man, it is highly unlikely that Kotondo Hasebe would have possessed the expertise for as formidable a task as excavating Longgushan. He would have required the cooperation and assistance of Pei, Jia, and other skilled Chinese workers, and it must have been apparent to all that such cooperation would not have been forthcoming. After World War II Hasebe did apparently work at physical anthropology, naming in 1947 a new species *Nippoanthropus akashiensis* on the basis of a pelvic bone found in 1931 in Akashi, Japan, that was destroyed during the firebombing of Tokyo. As evidence of his eclecticism, he also worked on determining the supposed racial affinities of some Japanese mummies, and collected and studied the bony remains of Japanese dogs.

Dr. A. B. D. Fortuyn had been a professor in the anatomy department of Peking Union Medical College with both Davidson Black and Franz Weidenreich. He last visited the medical school in July 1942, before leaving Beijing for London. At that time he had been summoned by a Dr. Matsuhashi, a Japanese epidemiologist who had taken over one of the physiology laboratories of the medical school. Dr. Matsuhashi wanted to know the whereabouts of the original *Sinanthropus* fossils. Fortuyn reports that he believed that the Peking Man fossils had made it out of Beijing and to the port city of Qinhuangdao. He relates that "this is known for certain, because the marine who personally was in charge of the boxes developed appendicitis after his temporary return to Peking. He was operated in the PUMC and then had a chance to pass this information to the attending doctor."[40] Fortuyn had believed until then that the Japanese had captured the fossils in Qinhuangdao and that they had been sent to Japan as a national treasure. He concluded after being interrogated by Matsuhashi that "the location of these fossils was at least not known to all Japanese who were interested in them."

Between July and August 1942, Pei wrote that "news suddenly came that the fossil was found in Tianjin (Tientsin) and the Japanese authorities were looking for someone to identify the authenticity of the object."[41] Claire Taschdjian was summoned by the Japanese to help, but when she arrived she was told to return home, as the fossil had been determined to have nothing to do with Peking Man. To this day the Chinese remain suspicious that the Japanese really did discover some or all of the Peking Man fossils.[42] Some believe that the fossils were sent to Japan. No further explanation was ever offered as to what object or objects had been mistaken for the Peking Man fossils.

Pei notes that at this point, August 1942, the Japanese suddenly dropped their investigation of Peking Man, and Hasebe, pleading lack of funds, abandoned plans to excavate at Zhoukoudian. He returned to Japan, taking with him some of the Zhoukoudian records and late Pleistocene fossil and archaeological collections. They were recovered in the Imperial Museum at the end of the war and sent back to China. But if the Peking Man fossils were among the relics, Kotondo Hasebe never had the opportunity to study or publish anything about them. A further curious fact is that on August 23, 1942, the *Peking Daily*, an English-language, pro-Japanese newspaper, published an article reporting that Hasebe and his assistant Takai had arrived in Beijing on August 19 and discovered that the Peking Man fossils had been removed from the safe at Peking Union Medical College.[43] Strangely, this fact had been discovered by Hasebe the day after Pearl Harbor, more than eight months earlier, as noted by Pei. Was this old news intentional "disinformation," disseminated to quell an unintended leak that the Peking Man fossils had been found? If the Peking Man fossils did get shipped to Japan, could they have been destroyed, like *Nippoanthropus*, in the massive bombing of Tokyo? If they were, they were not housed in the Imperial Museum, which escaped damage. Many other scenarios are possible, but it is clear that further historical research is needed to resolve outstanding questions on the Japanese side of the Peking Man fossil question.

The unfortunate fate of a large number of other, non-primate fossils captured by the Japanese at Peking Union Medical College is much clearer. Jia and Huang record and list the contents of 67 boxes of Zhoukoudian fossils and stone artifacts, 10 boxes of fossil reptiles from another site, and 30 boxes of publications that had been crated up and stored in Lockhart Hall when the Japanese took over the medical school in 1941.[44] In May 1942 the Japanese military police, at the height of the investigation into the Peking Man fossils, decided to move their headquarters into Peking Union Medical College. They ordered the fossils and books, which had up to then been undisturbed, thrown out. An eyewitness and former medical school employee, Deshan Han, recounted that "there were great numbers of bones on the ground, scattered and smashed," and that many books being thrown out and burned were retrieved by local residents and later sold to used book-sellers in Beijing. He himself picked up four pieces of fossil bone from the street that he later (in 1950) sent back to Dr. Yang.[45] Some fossils and casts were tossed into a storeroom at the medical school by the military police but were badly broken in the process. This type of treatment unfortunately would most likely have been the fate of the Peking Man fossils had they fallen into the hands of military personnel in wartime Beijing, even if those personnel were looking for them.

Fate of the Fossils, Science, and Responsibility

More ink has been spilled over the loss of the Peking Man fossils than any other historical topic in paleoanthropology, except perhaps the identity of the hoaxer of Piltdown Man. Despite all the interest, the historical re-search, and the hypotheses, there is still not a single reliable account of a sighting of the fossils since they were packed by Hu and Ji in 1941. It is human nature to speculate on the fate of the fossils, but there are impor-tant lessons to be learned as well.

Many believe that the Peking Man fossils still exist. Maybe they lie bur-ied somewhere, or stored away in some warehouse, or kept in hiding to be put up for ransom someday. The 2000 book by Ming-sheng Li and Nan Yue, likely reflective of mainstream Chinese sentiment, maintains that de-struction of the fossils is unlikely. These authors put a great deal of faith in the Americans' plan for evacuating the fossils. They suggest four possibili-ties for the fossils' fate: (1) that they were found by the Japanese and sent to Japan, where they remain; (2) that the Americans secretly changed the plan, and thus kept all Chinese in the dark, tricked the Japanese (who never did find the fossils), and likely brought the fossils to the United States; (3) that the fossils were buried somewhere, probably in China, by either the Americans or the Japanese; or (4) that the fossils were lost by either the Americans or the Japanese, in which case they could be virtually any-where and may yet be found.

On the other hand there is a less optimistic but much more realistic view of the fossils' fate. If one thinks of the mindless, wanton destruction that accompanies war, the desperate clawing for survival to which people are reduced during a war, the opportunistic scavenging that people revert to in wartime, and the history of destruction of hominid fossils during military operations, one forms a different view of the most probable fate of Peking Man. The documented loss of thousands of fossil and archaeological specimens from Longgushan when book-burning military police needed office space at Peking Union Medical College is a case in point. The bombing of museums in Berlin by Allied planes, destroying such specimens as Olduvai Hominid 1 and the Le Moustier Neandertal skeletal remains, and the Nazi destruction of the Predmostí Neandertal fossils in Czechoslovakia are other notorious examples. The misguided sense of relief expressed by Shapiro that the Zhoukoudian fossils were not obliterated by the atomic blasts at Hiroshima or Nagasaki ignores that 80 percent of Tokyo, where they may well have been, was destroyed by conventional Allied fire bombing. Misguided, too, because why should we assume that, with millions of human beings dying around them, the inanimate relics of Peking Man were somehow protected from destruction?

Our assessment is that the fossils of Peking Man are no longer intact and that they never left China. The bones were never seen again after they were packed at Peking Union Medical College. They did leave the college, but where they arrived even after their short journey, whether at the U.S. legation or the marine compound, is unknown. Had the fossils been transported out of China, their notoriety and the continuing interest in their fate over the years would likely have brought them to light. The chaos of China just after the attack on Pearl Harbor and the escalation of Japanese aggression almost certainly exposed the fossils to an unprotected setting in which they became no more than isolated "dragon bones," each with a definitive street value (the supposed worthlessness of the fossils was one of Houghton's biggest miscalculations). Like the scattered fossils in the street around the medical college picked up by locals in 1942, the Peking Man fossils, wherever they were scattered, were likely picked up as well. Ground up dragon bone drunk in tea is claimed by practitioners of traditional Chinese medicine to be, in addition to a cure for osteoporosis and male impotence, an excellent reliever of stress. And of that there was an abundance in the troubled times following the years of digging at Longgushan. One way or the other the bones were reclaimed by the dragon, the traditional protector of the Chinese earth and its treasures. One may hope that, as medicine, they helped innumerable Chinese cope with the ordeal of a world war.

This assessment of the fate of Peking Man is shared with Dr. Lucian W. Pye, professor of political science emeritus at the Massachusetts Institute of Technology. In 1945–46 Dr. Pye was an intelligence officer with the United States Marine Corps in China. He was assigned by the marines to find the Peking Man fossils, primarily for the "honor of the Marine Corps," which had been entrusted with their safekeeping at the beginning of the war. Dr. Pye and his men undertook an intensive investigation of warehouses in China in which the fossils may have been stored. The search was fruitless and Dr. Pye informed General MacArthur's command in Tokyo that a search in Japan might prove productive since there was no sign of the fossils in China. We have been unable to locate any independent records of this investigation, but Dr. Pye relates that two weeks later he received word from Japan that the fossils could not be located there either. At that time the investigation was dropped.[46]

Even if now destroyed, the bones of Peking Man permanently enriched science. Teilhard de Chardin was one who privately believed that the loss of the hominid fossils from Longgushan was not as catastrophic as it first seemed. In Claire Taschdjian's novel, Teilhard's character is quoted as saying, "The Sinanthrope has been dated, described, measured, x-rayed, drawn, photographed and cast in plaster down to the last fossa, crista and tubercle. . . . The loss is more a matter of sentiment than a true tragedy for science."[47] Taschdjian worked as Teilhard's secretary at the Institut de Géo-Biologie in Beijing from 1941 to 1946 and was well aware of his opinions (Teilhard also performed her marriage in 1946[48]). For scientists, much of the fossils' worth lay in the monographs, maps, and photographs that recorded their anatomy and geological context. As we show in the succeeding chapters in this book, much remains to be learned from the site of

Franz Weidenreich after leaving Beijing in mid-1941 in his office at the American Museum of Natural History in New York City. On the table in front of him are a cast of *Homo erectus* from Longgushan (*on pedestal*), a modern chimpanzee skull (*on left*), and a modern *Homo sapiens* skull (*on right*).

Longgushan and its fossils. The credit for this happy state of affairs is due to the untiring efforts of the Zhoukoudian scientists—Black, Yang, Teilhard, Pei, Weidenreich, and Jia—who documented the discoveries so well through a steady stream of publications, cast the specimens and sent the copies abroad, and photographed and mapped the site as it was excavated. But the loss of the original fossils does limit some types of direct research on the specimens. Could the loss of the specimens have been prevented, and are there any lessons for science? We believe the answer is yes in both cases.

Franz Weidenreich took on more than the anatomical description of the *Sinanthropus* fossils from Dragon Bone Hill when he stepped into Davidson Black's shoes. He inherited the curatorial responsibility of ensuring the safety of the collection. Ideally, Weidenreich the curator would have ensured the safety of the fossils before he left China, either by carrying them to safety himself or ensuring their removal to a safe location in China. But as we have seen, he did not have a choice in the matter. The record does show that Weidenreich pushed successfully for action by the Rockefeller Foundation as soon as he arrived in New York.

As the government official in charge of the Peking Man fossils, Weng Wen-hao had an even more onerous curatorial duty. His was to vouchsafe the Longusshan specimens for China. But what is the appropriate course of action when the control of a country is in dispute? In hindsight, had Weng requested the Peking Man fossils from the medical school as soon as the Nationalist government vacated Beijing, and then simply buried them somewhere until after the war, the fossils' whereabouts would have at least been known. The Peking Man fossils were widely known as a Chinese national treasure and they would likely have been kept intact by any Chinese faction into whose hands they fell. It is possibile that Nationalist Chinese agents succeeded in locating the Peking Man fossils before the Japanese, kept the discovery secret, and transported them to Taiwan at the end of the war. So far as we are aware, a concerted search for the Peking Man fossils among relics that might have been taken from the mainland to Taiwan by the Nationalist government has never been undertaken. Weng's strategy, however, was to pursue the only avenue that he thought feasible— evacuation of Peking Man by the U.S. government and the Rockefeller Foundation. His plan might have worked had it been put into operation earlier in 1941.

The motives of the Imperial Japanese Army and the Japanese scientists in China attempting to loot the Peking Man fossils seem to have been grounded in a sort of Napoleonic cultural imperialism. The Japanese wanted to acquire the Peking Man fossils not only because the fossils had international scientific importance and represented their ancient mainland ancestors, but because the Chinese considered them national treasures. By capturing Peking Man, the Japanese could seal their domination of the Chinese, whose country they now controlled militarily. However, the importance of scientific specimens such as Peking Man transcends nationalism, if this indeed was at the root of the Japanese interest in the fossils. Our reading of the evidence is that the Japanese had no more success finding *Sinanthropus* in China than they did finding the bulk of the *Pithecanthropus* fossils, which had been well hidden in Java by Ralph von Koenigswald.

Dr. Henry Houghton occupies a unique place in the history of Peking Man. Originally and to the end, he wanted his medical school to have no part in ancient caves and dusty fossils. But the enthusiastic conspiracy of his faculty, young Chinese researchers, and the international scientific community had just been too much for him to withstand. He had reluctantly agreed to allow the medical school to serve as the focus for one of the largest paleontological projects in history—a purpose in his opinion far removed from medicine. Yet it fell to Houghton, when war loomed on the horizon, to ensure the safety of the invaluable fossils. One would have thought that he would have welcomed the opportunity to rid himself of Peking Man as soon as possible. Instead, he opposed the removal of the fossils from the medical school on the basis that fragile political relations in Japanese-occupied Beijing might be compromised. When it became apparent that those relations were irreparable, he should have acted expeditiously. But Houghton's concerns were never for the scientific importance of the Longgushan fossils, and on December 8, 1941, time ran out. Houghton had finally acted, but the Peking Man fossils were caught somewhere in the swirling chaos of world war. Houghton, along with Controller Bowen, as the last of the American administration of the Peking Union Medical College left in Beijing, were interned by the Japanese on December 8, 1941. They remained imprisoned for the duration of the war. The safety of the fossils had been Houghton's responsibility, but unlike Ralph von Koenigswald, who dug up his fossils when he was released from a Japanese prison in Java at the end of the war, Houghton would have had no idea where the Longgushan fossils might be, even if he had cared to look.

The disappearance of the Longgushan fossils represents the single greatest loss of original data in the history of paleoanthropology. There are those who still search for them. We wish them well, for there is a slight chance that the Peking Man fossils escaped the massive forces of destruction run amok in World War II. Unfortunately, it is most likely that, as soon as they left their safe haven in the rarified environment of a scientific laboratory, they were transformed back into dragon bones. As dragon bones once more, they probably were sold as valuable commodities and then most likely consumed as medicine. We strongly suspect that, like the bones of Dr. Davidson Black, whose grave was razed during the postwar reconstruction of Beijing,[49] the bones of the hominids that he helped discover have now been irretrievably commingled with the earth of China.

CHAPTER 3

Giants and Genes:
Changing Views of Peking Man's
Evolutionary Significance

Today scientists who study human evolution use an increasingly multidisciplinary approach to test their hypotheses. Nevertheless, the theories of fossil-based evolutionists and those of molecular-based evolutionists have resulted in a lively intellectual competition. This clash of mindsets and theoretical persuasions first became evident in the interpretations of *Sinanthropus*. Franz Weidenreich was a pivotal figure.

In this chapter we will look at the twists and turns of the paleoanthropological interpretation of Peking Man—the first, largely fanciful and hopeful pronouncements; the more mature hypotheses based on the remarkably complete fossil remains; the comparative studies of Zhoukoudian with other sites and hominids from around the globe; and finally the understanding of Peking Man after all the pre–World War II fossils had been pulled out of the ground. The ideas evolved as the data accumulated from this remarkable site and, as we shall see, they led us to our modern conception of this unique human species.

Anatomy of Peking Man Revealed

Franz Weidenreich was a methodical and tidy man, a natty dresser who worked very hard and kept a keen eye on the budget of his laboratory and the excavations at Longgushan. His energy seems to have sprung from a deep-seated passion for his work rather than a dedicated professional work ethic alone. He plunged into the task of describing the new *Sinanthropus* fossils. Once, when he was presented with some newly discovered fossil hominid skull fragments from Longgushan by excavator Lanpo Jia, his

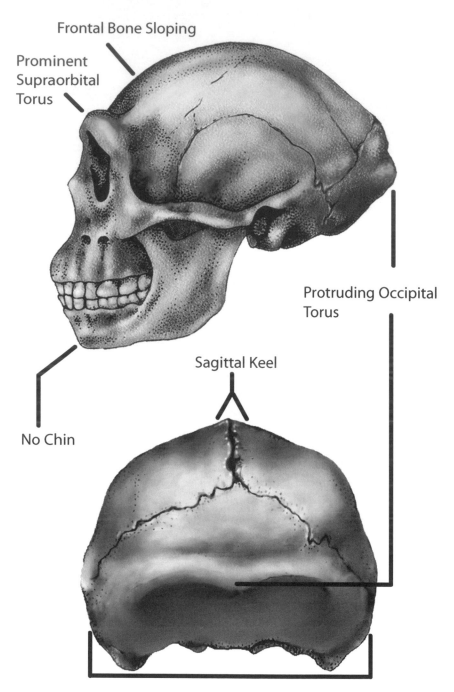

Frontal Bone Sloping

Prominent Supraorbital Torus

Protruding Occipital Torus

No Chin

Sagittal Keel

Cranial Width Broadest at Base

hands shook with excitement as he turned the specimens over and over, and all he could say was "wonderful, wonderful."[1]

Weidenreich was impressed most of all with the profile of the Peking Man skull. Whereas our modern skull is large and globular, balancing upon our neck like an inflated bony balloon, Peking Man's skull resembles the carapace of a turtle, low and crouching, massive and rimmed with thick bone. Whereas the brains contained within a modern *Homo sapiens* skull are enough to fill three-quarters of a half-gallon milk container, the brain of *Sinanthropus* would almost fit in a quart.[2] Weidenreich coupled these two observations in an explanatory framework that has stood almost intact until the present day. In 1939 he wrote that the "extraordinary primitiveness" of the *Sinanthropus* skull "must be considered a consequence of the smallness of the cranial capacity."[3]

Weidenreich believed that as the brain expanded from its primitive apelike state, a number of changes in cranial anatomy occurred. Without knowing what many of the unique anatomical traits of *Sinanthropus* meant, he struggled to interpret them in an evolutionary progression from ape to modern human. He compared the supraorbital tori—the large bony ridges projecting from above the eye sockets—to similar large structures in apes. He measured the length and breadth of the braincase and compared its long and low shape in *Sinanthropus* to that of ape skulls. He measured the position of the foramen magnum, the hole in the base of the skull through which the spinal cord passes, and noted that it was located more to the back of the head, again, more apelike than in modern humans. Where *Sinanthropus* differed from apes, Weidenreich attributed the change to progressive evolution toward *Homo sapiens*. For example, he noted that what little was known of the *Sinanthropus* facial skeleton showed a reduction in its overall size, and the canine teeth of *Sinanthropus* were reduced in size and humanlike in form.

Weidenreich was too good an anatomist not to realize that some aspects of the strange skull form that he had discovered and described in *Sinanthropus* did not make sense in terms of brain size expansion alone. He observed and named a new structure—the "torus occipitalis"—a linear bony thickening that ran from side to side at the back of the skull. Apes do not have such a structure, although they may have a sharply defined shelf of bone at the back of their heads to which the strong neck muscles attach.

Facing page

Anatomy of the *Homo erectus* skull as reconstructed by Franz Weidenreich. The skull has a low and elongated form as compared with modern humans. A massive buildup of bone above the eyes, the supraorbital torus, and a thickening of bone at the back of the head, the occipital torus, magnify the effect. The top of the skull shows a characteristic, rounded keel of bone down its middle, extending to flattened bone surfaces on either side.

The *Sinanthropus torus occipitalis* was round, smooth, and completely different. It is located much higher on the back of the skull than the attachment of the relatively small neck muscles. Therefore, the *torus occipitalis* of *Sinanthropus* must be for something other than muscle attachment. Puzzled by the anatomy, Weidenreich sidestepped the issue. Avoiding a functional interpretation of the *torus occipitalis*, he used this trait to relate *Sinanthropus* to the later Neandertals, whom he considered to be ancestral to modern humans. Neandertals have a small midline remnant of the *torus occipitalis*, an expansion on the occipital bone known as an "occipital bun" (widely referred to by the French term "chignon").

Weidenreich explained other confusing anatomy of *Sinanthropus* as either irrelevant to phylogenetic interpretation or as variations that might indicate some continuity across species boundaries to modern humankind. In the first category he placed the rounded crest of bone that in *Sinanthropus* runs from the middle of the forehead back along the crown of the head. Like the *torus occipitalis*, this structure is a low, rounded local thickening of bone, very unlike the thin, ridge-like midline "sagittal crest" that serves for the attachment of the large temporalis chewing muscles of apes. The "temporal lines" to which *Sinanthropus* temporalis muscles attached are clearly preserved and located far below this structure. We now term it a "sagittal keel." Weidenreich sidestepped the anatomical meaning of the sagittal keel, which was as puzzling as the occipital torus, and instead considered a "strong sagittal crest" in *Sinanthropus* one of several features which, "though irrelevant from the phylogenetic standpoint, . . . nevertheless occur in exactly the same manner and degree of formation in recent mankind."[4] He noted that similar bony excrescences could be seen on some modern Asian, Australian, and Tasmanian human skulls.

The most well-known example of a trait that Weidenreich adduced to indicate evolutionary continuity from *Sinanthropus* to *Homo sapiens* was the shovel-shaped incisor. American anthropologist Aleš Hrdlička had introduced the term in 1920 and noted it occurred frequently in Asians and American Indians.[5] The type of "shovel" to which he compared these teeth is one with a blade that has upturned edges, much like an old-fashioned coal shovel. If you happen to have shovel-shaped incisors you can feel the ridges with your tongue on the back of your front teeth. Weidenreich documented shovel-shaped incisors in Longgushan *Sinanthropus,* whose four upper front incisors and lower second incisors generally show the trait. Modern Asians and American Indians (who have a recent Asian ancestry) have by far the highest incidence of shovel-shaped incisors among modern people, a fact that Weidenreich used to infer a *Sinanthropus* ancestry for *Homo sapiens*.

By 1943 Franz Weidenreich had done an excellent job of fully describing to the scientific world all of his careful observations on the Chinese fossils from Dragon Bone Hill. Regardless of the exact evolutionary meaning of the anatomy of *Sinanthropus*, it was certainly different and it was certainly old. All paleoanthropologists recognized it as important and in need of theoretical explanation. The place to start to look for answers was in comparisons with previous discoveries and what was already known. To whom or what was Peking Man related?

The Kinship of Java Man and Peking Man Recognized

Davidson Black's ideas on *Sinanthropus* had begun to form as soon as he laid eyes on the first fossil hominid teeth from Longgushan. His initial conception of what the Longgushan hominid would look like was a Piltdownesque modern human skull and ape jaw. But he was not alone in this misconception.

Sir Grafton Elliot Smith was Black's mentor, consistent correspondent, and most ardent supporter in England. Soon after the 1929 skull was found, Black invited Smith to China, and he came the following year. Smith studied the new skull discoveries with Black at his side, and his 1931 paper reviewing the *Sinanthropus* discoveries from China was an influential endorsement of the Longgushan research from one of the world's most recognized authorities. Smith grappled with the issue of relatedness between the new Chinese fossils and those from Piltdown, on the one hand, and from Java, on the other. He wrote, "Just as the finding of the jaws in 1928 suggested the possibility of some kinship with the Piltdown man, the skull found in 1929 caused opinion to swing in the other direction and suggested a nearer kinship with *Pithecanthropus*. In 1930, however . . . the braincase was revealed with a curious blend of characters hitherto regarded as distinctive, some of them of *Pithecanthropus* and others of *Eoanthropus* [Piltdown]."[6] Elliot Smith's own drawings comparing the new Chinese skull and Piltdown are still instructive today, even though we now know that Piltdown was a forgery, a modern *Homo sapiens* with pathologically thickened skull bones associated with a modified orangutan lower jaw. When Davidson Black died at his work bench among his fossils in 1934, he had recently recognized the inescapable similarity between the skull forms of *Sinanthropus* and Javan *Pithecanthropus*, but his opinion on the evolutionary relationships of *Sinanthropus* was still very similar to that published by Elliot Smith four years earlier.

When Franz Weidenreich came onto the scene, he had a different intellectual background, a different set of assumptions about human evolution, and a growing collection of fossils from which to generate and test evolutionary hypotheses. For Weidenreich, Piltdown Man, so enthusiastically embraced by British and North American paleoanthropologists, was not a major problem. He, like Gustav Schwalbe, his former professor and mentor at Strassburg, gave little credence to the Piltdown discovery. They thought that the jaw was some sort of fossil ape similar to an orangutan that had lived in England, but the skull they recognized as essentially modern human, that is, non-Neandertal in its anatomy. Weidenreich called the Piltdown fabrication a "chimera"—in reference to the fantastic hybrid beast of Greek mythology—even though the fraudulent nature of these remains was not fully proven until 1953, five years after his death.

Weidenreich's opinion of Piltdown allowed him to see the close anatomical similarities between his Chinese fossils and those found in Java three decades before by Dutch anatomist Eugene Dubois. Dubois had discovered only a single skullcap of what he had named *Pithecanthropus erectus*, a specimen that Weidenreich had studied while still in Germany.[7] More specimens from Java were badly needed in order to test the idea that *Pithecanthropus* and *Sinanthropus* were anatomically similar and therefore closely related.

Hominid fossils in Java were discovered by an adventurous young German who had been interested in fossils since childhood and had rather romantically fashioned himself a latter-day Dubois. G. H. R. ("Ralph") von Koenigswald, a 28-year-old paleontology Ph.D. who was two years out of the University of Munich (where Weidenreich had also studied as an undergraduate many years before), went to Java in 1930 to work for the Dutch Geological Survey.[8] His primary job was to do geological mapping, but fossils were a natural part of the stratigraphic phenomena that needed to be described. Von Koenigswald immediately went to Dubois's old sites at Trinil, writing his first paper on hominids in Dutch in 1931.[9] The first word of the paper's title is "*Sinanthropus*," underlining von Koenigswald's early appreciation of a connection between China and the fossils that he was discovering in Java. He had also begun snooping around Chinese apothecary shops in Java, buying "dragon bone" fossils. In the tradition of Haberer and Schlosser, one of his former professors at Munich, he published papers on these discoveries.[10] He was in Java when, between 1931 and 1933, 11 fossilized late Pleistocene human skulls were found by a Dutch Geological Survey team at a place called Ngandong along the Solo River. They were announced by Dutch geologist Cornelius ter Haar and, despite the fact that the Ngandong discoveries were relatively young geologically, the significant and renewed potential of Java for fossil hominid studies had become apparent.

Following the discovery of the Ngandong skulls, von Koenigswald received a grant from the Carnegie Institute of Washington (D.C.) in 1934 to work full-time on prospecting for hominids independently of the Dutch Geological Survey. He hired a number of local Javanese to help him, and this approach was to prove successful. Although the Javanese did not share the Chinese interest in dragon bones, their constant tilling of the land occasionally turned up fossils. As soon as von Koenigswald's interests became known, collectors began bringing fossils to him. The technique was not without its drawbacks, however. Sometimes collectors would intentionally break an intact fossil in an effort to exact a higher price, that is, for "two" fossils. And it was usually impossible to find out the exact location of a discovery because the collector wanted to keep this information to himself. But with persistence, and more money, von Koenigswald succeeded in making a number of important fossil discoveries. And, importantly, he was eventually able to ascertain where the fossils came from so that earth scientists could then determine the geological age of the fossils.

In April 1934 von Koenigswald discovered his first early Pleistocene hominid fossil, a mandible from a site called Sangiran. It was the first early hominid discovery in Java since Dubois's discoveries before the turn of the century. Then in 1936 a much older single skull of a juvenile hominid was discovered at a place called Mojokerto. His publications appeared in 1936 and 1937.[11] Somehow Franz Weidenreich managed to obtain copies of the papers in Beijing and avidly read of von Koenigswald's discoveries, even as the Sino–Japanese War engulfed the region and halted all excavation in China. Von Koenigswald discovered an even more complete hominid skull in 1938, of the same geological age and with the same anatomy as Dubois's original *Pithecanthropus erectus* skull found some 45 years before.[12] Weidenreich lost no time contacting von Koenigswald in Java and visiting him the same year. He was to write that this newly discovered skull "resembled Dubois' original skullcap as one egg does another."[13]

It was natural that Weidenreich and von Koenigswald would get together. Both were German anthropologists[14] in the Far East and both were trying to work out the evolutionary relationships of the unusual fossil hominids that they were studying. Weidenreich badly needed more hominid material in order to understand the missing pieces of the Longgushan puzzle, and the excavations in China had been halted by the war. Von Koenigswald, on the other hand, was a young and unproved paleoanthropologist who needed an experienced ally. He had already been compromised by none other than Dubois himself who, in 1936, as editor of the leading Dutch scientific journal, had changed the proofs of an article in which von Koenigswald had proposed a new species name, *Pithecanthropus modjokertensis*. Dubois changed it without von Koenigswald's knowledge to *Homo modjokertensis*. Von Koenigswald, who had earlier greatly admired Dubois, never forgave him. Forty years later, when one of us discussed this matter with him, von Koenigswald was still fuming.[15] Dubois would never have dared to do that to someone of Weidenreich's stature. When von Koenigswald discovered a second hominid skull in Java, he published it—his first paper in English—with Weidenreich in the British journal *Nature*.[16]

Weidenreich was becoming increasingly convinced that there was a close connection between Javan *Pithecanthropus* and Chinese *Sinanthropus*. In 1939 he wrote that "*Pithecanthropus* is a genuine hominid [versus a giant gibbon as then believed by Dubois] of about the same general stage of evolution as *Sinanthropus*."[17] But he needed more anatomical proof, and the fossils were coming out of the Pleistocene sediments of Java. The year after Weidenreich's 1938 trip to Java, von Koenigswald visited him in Beijing, bringing along more newly discovered hominid fossils. The two men decided to join forces in interpreting their respective fossils.

In 1939 von Koenigswald and Weidenreich published their definitive position on the issue of the biological connection between Chinese *Sinanthropus* and Javan *Pithecanthropus*.[18] By directly comparing their fossils they were able to demonstrate that the two populations of ancient hominids were very closely related, if not identical. Such a collaborative effort marked a unique event in paleoanthropology. With this connection established between Chinese and Javan hominids, Weidenreich had a larger sample from which to draw evolutionary conclusions. The Javan fossils were to become uniquely important in Weidenreich's interpretations of the Chinese hominids from Longgushan.

A Pleistocene Land of Giants: Robust *Pithecanthropus*, *Meganthropus*, and Gigantic Apes as Ancestors

Without needing the permission of a cumbersome bureaucracy spanning two continents, as did Weidenreich, Ralph von Koenigswald simply took his Javan fossils, which included the Ngandong specimens, and buried them in his garden when calamity threatened. In 1941 he was captured by the invading Japanese army and spent most of the war interned or in a prison camp in Java. Only one of the fossils, Ngandong Skull XI, was confiscated by the Japanese. After the war it was discovered in the emperor's household museum in Tokyo by a U.S. army officer who had studied anthropology in college,[19] and who later returned it to von Koenigswald. When the war ended in 1945, von Koenigswald dug up his fossils and was soon on a ship to New York with them.

In the years following his 1941 departure from China, and while von Koenigswald was incommunicado in Japanese-held Java, Weidenreich labored over his magnum opus, his final description and interpretation of the Longgushan fossils—"The Skull of *Sinanthropus pekinensis*, A Comparative Study of a Primitive Hominid Skull."[20] The salient anatomical features of this species had been known in general terms for years, and many of the observations were the same as for the first skull and had been published by Davidson Black. But Weidenreich wanted to know *why* the *Sinanthropus* skull looked the way it did.

Weidenreich's explanation for the profile of the skull relied largely on the small brain size of *Sinanthropus*. Some of the other aspects of the skull, such as the projecting browridges, could plausibly be related to the small brain size. Weidenreich also thought that standing upright on two legs had transformed the *Sinanthropus* skull. He thought that as the human skull became adapted to sitting atop the spine, it had folded, bending at a crease running from ear to ear under the skull, thereby bringing the face down and under the skull, flexing its basicranium. The face of *Sinanthropus* had become shorter and less projecting than the faces of apes in this process called "basicranial flexion." Weidenreich's final word on his explanation was a volume written as a *Memoir of the American Philosophical Society* entitled "The Brain and Its Role in the Phylogenetic Transformation of the Skull."[21] Here he argued that as the brain increased in size it caused an underfolding of the face, a reduction in forward projection of the face ("prognathism"), and resulted in a more globular head balanced upon the vertebral column. To bolster his point, Weidenreich used examples from the skull anatomy of domestic dogs, which showed a trend toward reduction of the snout with increased brain size.

Most anatomists and anthropologists were convinced by Weidenreich's arguments—to a point. The problem of the thick skull bones and the massive, rounded tori of bone around the *Sinanthropus* skull continued to bedevil a straightforward interpretation of the evolutionary transformation of small-brained, apelike skull to large-brained, humanlike skull. Neither modern humans nor apes have this skull anatomy. And if large-brained primates like people do have thick skulls, and smaller-brained primates like apes do not have thick skulls, what then causes *Sinanthropus* to have a thick skull? Logically, it must be something other than brain size. The Javan fossils were to provide Weidenreich with what he thought was his answer.

By 1941 von Koenigswald had discovered and published on fossils from Java that could be interpreted in a size-graded series—from large to small. Weidenreich was able to construct what he considered a gigantic ancestry for hominids. From a large jaw that von Koenigswald had named *Meganthropus palaeojavicus* ("huge man of ancient Java"); to an upper jaw with teeth that von Koenigswald had named *Pithecanthropus robustus* ("robust ape-man"); to *Pithecanthropus erectus*, the smallest of the Javan group; to, finally, the most gracile of his hypothesized lineage, *Sinanthropus pekinensis* from Longgushan, this sequence explained skull thickness by a heritage of huge body size. It was a bold hypothesis and one that caused Weidenreich's colleagues to raise their eyebrows.[22] Yet it explained what was a very important anatomical peculiarity of the *Sinanthropus* skull. If *Sinanthropus* had descended from giants, then the non-modern-human and non-modern-ape aspects of his skull anatomy could be explained as a primitive retention. Weidenreich published his gigantism theory in the 1946 book *Apes, Giants, and Man*, and in a joint paper with von Koenigswald.

In addition to its novelty, Weidenreich's theory had another problem. Where were the giant apes from which giant hominids evolved? The living Asian apes, the orangutan and the gibbon, were too small to be convincing descendants of some gigantic ancestor. The only living ape that might fit the bill was the gorilla, but gorillas live only in central and western Africa, a long way from China. However, a gorilla-sized fossil ape had been discovered and named by von Koenigswald, who had discovered isolated teeth of such a creature in the dragon bones that he had bought in southern China. He had

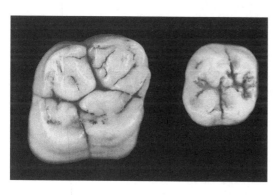

named this extinct Chinese ape *Gigantopithecus blacki*, in honor of Davidson Black.[23] Weidenreich seized on this undated and poorly known ape as the progenitor of his giant hominid lineage, even suggesting that it should be renamed *Giganthropus* ("giant human").

Weidenreich's gigantism theory of human evolution is today largely forgotten, having been disproved by a rising tide of discovery of earlier fossil forbears from Africa and Asia. We now know that the human species descended not from giants but from pygmy-sized early hominids in Africa millions of years before any of them ever ventured into Asia. Weidenreich hypothesized that India would be the place that ultimate human origins would be discovered. In this he was

Homo erectus compared to *Gigantopithecus*. Franz Weidenreich proposed a theory in 1945, not supported by subsequent discoveries, in which giant hominoids such as *Gigantopithecus* gave rise to early humans. Figured are two upper molar teeth bought in Chinese apothecaries by G. H. R. von Koenigswald: *Gigantopithecus blacki* (specimen 3), recovered in Canton, China, in 1935 (*left*), and *Homo erectus* (type specimen of *Sinanthropus officinalis*), recovered in Hong Kong in 1935 (*right*).

also mistaken. Indo–Pakistan is now known as having been a locus of ape evolution 8 to 12 million years ago that gave rise to ancestors of the orangutan and to *Gigantopithecus*, still a poorly known ape but certainly not a hominid. But Weidenreich's gigantism theory explained one important aspect of *Sinanthropus* anatomy that remained unaccounted for when his theory was tossed out. Why did the remarkably thick and robust skull of *Sinanthropus* evolve? And if not from the pachyostotic skull form of giant ancestors, then how? We propose a new theory in chapter 4.

Weidenreich, Multiregionalism, and the Dawning Realization of *Homo erectus* as a Zoological Species

At the same time that paleoanthropologists were sifting through their data and refining the interpretation of Dragon Bone Hill and its hominids, pressure was mounting for anthropology to conform to the tenets of modern biology. Genetics and population biology, integrated with Darwin's theory of natural selection, had made evolutionary biology a new "synthetic" discipline. Old names, genera like *Sinanthropus* and *Pithecanthropus*, were criticized as conferring too much distinctiveness on populations of hominids that in the past were probably all members of one zoological species. Instead, the new biologists suggested that most, or even all, known fossils of early hominids would fit into several species all within one genus, our own, the genus *Homo*.

Paleontologists in general were wary of the new biology. They spent much of their research time and major portions of their careers attempting to show how their fossils were different from all previously discovered fossils, and thus were valid new species. Few paleontologists were interested in naming their newly found fossils the same as someone else's fossils. Similarities between fossils tended to be overlooked. Therefore, Weidenreich's recognition of anatomical similarity and thus zoological relatedness between Chinese *Sinanthropus* and Javan *Pithecanthropus* was greeted with acclaim by the new biologists, who urged him to name them both *Homo*, in accordance with the tenets of their new synthesis. But Weidenreich never agreed to use the new nomenclature even though he agreed with many of the new biologists' theoretical points. Perhaps he had used the old names too long and they held too many memories for an old man to give them up. He said that he would leave that to others.

Even if Weidenreich's naming conventions for fossil hominids remained old-fashioned, his grand theory of human evolution was influenced strongly by the new biologists. It contained a model that has been influential to the present day. He proposed what has become known as "multiregionalism"[24]—the idea that there was genetic interchange among populations at any one time in the past (as there is today), and thus a greater degree of similarity among individuals within a region (within one "race," in his terminology) than between individuals from different regions. A corollary of this idea was that there could be evolutionary continuity of regional anatomical differences through time, even across species boundaries. This concept was important for explaining aspects of the Longgushan hominids' anatomy—shovel-shaped incisors, for instance, and their presence in modern Chinese people.

Weidenreich drew an interconnecting matrix of interbreeding populations at successive time periods in the past to explain his concept. His

	Phase	Horizontal Differentiations			
		1 Australian Group	2 Mongolian Group	3 African Group	4 Eurasian Group
Neoanthropinae	I Hos	Australian group	Mongolian group	South African group	Eurasian group
Neoanthropinae	IX Hof	Wadjak group (Java)	Choukoutien (Upper Cave)	Boskop group	CroMagnon group (W. Europe)
Paleoanthropinae	VIII Pae				Skhūl group (Palestine)
Paleoanthropinae	VII Pan				Tabun group (Palestine)
Paleoanthropinae	VI Par			Paleoanthic rhodesiensis	
Archanthropinae	V Pis	Pithecanthropus soloensis			
Archanthropinae	IV Pie	Pithecanthropus erectus	Sinanthropus pekinensis		
Archanthropinae	III Pir	Pithecanthropus robustus			
Archanthropinae	II Meg	Meganthropus			
Archanthropinae	I Gig		Gigantopithecus		

Weidenreich's (1945) trellis of multiregional evolution in the hominid family. Population networks are connected by the exchange of genes. This model included Weidenreich's outmoded gigantic theory of human origins, but it is important because it also incorporated early ideas about population genetics in human evolution. Multiregional evolution entailed significant vertical gene transmission from ancestors to descendants, within four regional groupings (horizontal differentiations), with less significant gene exchange between these groups, signified by the heavy lines between them. Archaic taxonomic terms used by Weidenreich are Archanthropinae (minus *Gigantopithecus*, largely synonymous with *Homo erectus*), Paleoanthropinae (*Homo heidelbergensis* and Neandertals), and Neoanthropinae (anatomically modern *Homo sapiens*).

meaning was clearly that species would evolve as interbreeding units across a broad front, even though some traits might survive preferentially in one population within one region. A colleague, Carleton Coon from Harvard University, misunderstood this aspect of Weidenreich's model and proposed in his book, *The Origin of Races*, five separate lineages of human races going back to the time of *Homo erectus* (obviously untenable for a modern species with fully interfertile populations). Biological anthropologists later corrected this misconception of Weidenreich's model and it became an important foundation for the modern multiregional interpretation of human origins and the single-species hypothesis.

Weidenreich's model explained the anatomy that he had observed in the Chinese and Javan hominids and also set the stage for a revolution in taxonomy that was to sweep anthropology in the 1950s. One of the new biologists, an ornithologist by the name of Ernst Mayr, made a proposal that if all early hominid populations at any one time in the past freely interbred, as Weidenreich had suggested, then by definition they had to be one species. He proposed that all fossil hominids then known be referred to by the single genus name *Homo*. Thus was born the single-species hypothesis, a powerful model that endured until the late 1970s when fossil discoveries in Africa disproved it, at least for the early part of the hominid fossil record. In the meantime, the panoply of Latin and Greek binomials was pared down substantially as fossils were compared and classified with categories that attempted to recognize their true biological relatedness. Weidenreich and von Koenigswald had unwittingly promoted this undertaking when they had concluded that Chinese and Javan fossil hominids were closely related. Now, using Mayr's new rules, both the names *Pithecanthropus* and *Sinanthropus* were relegated to the trash bin of paleoanthropology. Henceforth the Longgushan fossils and their Javan conspecifics became known simply as *Homo erectus*.

Franz Weidenreich died in 1948 bequeathing a wealth of anatomical detail and reasoned interpretation of hominid evolution to a generation of biological anthropologists ill equipped to deal with either. The leading academic biological anthropologist of the day was Ernest Hooton of Harvard University, an erudite and witty classical scholar whose extent of training in the anatomical and biological bases of human evolution consisted of brushing shoulders with Sir Arthur Keith in England. Nevertheless, he was responsible for training much of the next generation of biological anthropologists in America. Sherwood Washburn, one of Hooton's former students, edited Weidenreich's last papers and published them as a memorial volume.[25] Only one young undergraduate student of Washburn's at the University of Chicago, the future paleoanthropologist F. Clark Howell, briefly studied with Weidenreich at the American Museum of Natural

History. With this exception, the rich tradition in anatomy and evolutionary anthropology from which Weidenreich had emerged in Germany, and which he had so brilliantly carried into China, was truncated. Today the Franz Weidenreich Institute (founded by Weidenreich in 1928) at Johannes Goethe University in Frankfurt represents an attempt to recapture some of this tradition, which, along with the Peking Man fossils, was an unfortunate casualty of World War II. Reiner Protsch, a German trained at UCLA in the 1960s by another Hooton student, Joseph Birdsell, returned to Germany to resuscitate Weidenreich's old institute.

The Scientific Fate of *Homo erectus:*
The Muddle in the Middle

By 1946 the world had realized that the famed Peking Man fossils had been lost during World War II. Their scientific memory was kept alive by the comprehensive publications of Weidenreich. But as masterful as these works were, Weidenreich's monographs and papers became much like a requiem to the lost fossils. Not only were the originals no longer available for study and comparison by other scientists but there was no hope of excavating again at Longgushan to obtain new fossils. While Weidenreich finished the last of his publications in New York, civil war raged on in China between the Nationalists and the Communists. Eventually the Communists, under Mao Tse-tung, would prevail, and in October 1949 the People's Republic of China was proclaimed. By then Weidenreich, who had harbored hopes of going back to China to resume the work at Longgushan, had died.

China remained politically and scientifically cut off from the West for many years, as other wars, in Korea and Vietnam, ravaged Asia. Although this lack of scientific communication clearly contributed to a decline of research on *Homo erectus*, another factor was much more important in causing the fading of scientific awareness of the species. The tide of discovery in Africa of earlier and more primitive hominids known as australopithecines, and the development of new methods of accurately dating them, occupied most paleoanthropological researchers in the 1950s and subsequent decades.[26] Early hominid fossils coming out of South and East Africa eclipsed *Homo erectus* and drew attention away from Asia as a center of human evolution. *Homo erectus*, while not forgotten, was unceremoniously shoved out of the limelight.

Homo erectus then turned up unexpectedly in the African country of Tanzania, uncovered at the famous site of Olduvai Gorge by Kenyan-British paleoanthropologist Louis Leakey in 1960. The skullcap that was found eroding out of Bed II could have easily fit into the series of skulls found at Longgushan, except that Longgushan was two continents away and of younger geological age. Estimates of the age of the skull, called Olduvai Hominid 9, hovered around 1.4 million years. Leakey, who was avowedly looking for the direct ancestor of modern *Homo sapiens* in Africa, seemed embarrassed to have found a species so closely associated with Asia and one that he considered completely off the line of evolution to more advanced humans. In 1963 the German paleoanthropologist Gerhard Heberer named the new fossil *Homo leakeyi*, but Leakey barely acknowledged the honor. Even he recognized that the skull was closely allied to fossils published by Dubois, Black, and Weidenreich and now widely assigned to *Homo erectus*. Leakey was much more enamored of the earlier hominids that he and his team had discovered at Olduvai that became known as *Australopithecus* (originally *Zinjanthropus*) *boisei* (nicknamed "Dear Boy") and *Homo habilis* (lovingly termed "Olduvai George," "Cinderella," "Jonny's Boy," and "Twiggy"). Olduvai Hominid 9 (OH 9), an out-of-place Asian black sheep, never got a cute nickname. He was pushed to the back of the museum shelf and largely forgotten. Leakey regarded OH 9 as a specialized if not aberrant hominid that was not closely related to the human lineage.

Leakey's opinion of African *Homo erectus* was never mainstream, but he did have some basis for his ideas. Despite the fact that *Homo erectus* fell between *Homo sapiens* and *Homo habilis* in brain size, it had the thick skull bones and strange cranial tori that neither the earlier nor the later species possessed. Leakey chose to draw a line on his evolutionary tree directly from *Homo habilis* to *Homo sapiens*, bypassing *Homo erectus*. Only recently have further fossil discoveries in Africa, Asia, and Europe definitively

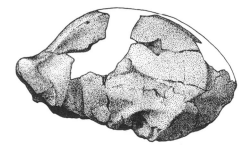 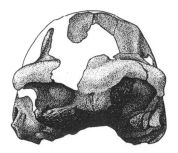

The black sheep of Louis Leakey's fossil hominid discoveries from Olduvai Gorge, Olduvai Hominid 9—a *Homo erectus* skullcap from Upper Bed II, dating to approximately 1.4 million years ago. *Left:* side view. *Right:* front view. Scales are 1 cm.

refuted this hypothesis by documenting fossil intermediates connecting *Homo erectus* with both earlier and later species.[27]

The discovery in Africa of *Homo erectus* was also ultimately to disprove Weidenreich's hypothesis of the giant ancestry of *Homo sapiens*. Hundreds of hominid fossils from Africa, and even fossilized footprints, now document that our earliest hominid ancestors were small-bodied. *Homo erectus*, far from being the smallest of an evolving human lineage, is now known to have been the first hominid of virtually modern human body size and proportions. Weidenreich's early ideas, advanced to explain the unusually thick cranial vault and other anatomical peculiarities of *Homo erectus*, fell prey to a more complete fossil record found in better-dated contexts. Yet the time between the demise of *Homo habilis,* about 1.6 million years ago, and the appearance of *Homo heidelbergensis,* now put at approximately 600,000 years ago, has been aptly termed the "muddle in the middle."[28] This one million-year period of time was when hominids left Africa and populated most of the Old World. *Homo erectus* was the species responsible and we are only now learning how this amazing event happened.

Bones and Genes: Apples and Oranges or Peas and Carrots?

Some say that bones and genes, when studied in our evolutionary biology, are like apples and oranges—sort of the same but not really, so they should be considered separately and individually. Others believe that skeletal anatomy of fossils ("bones") and data from biomolecular analyses ("genes") go together like peas and carrots—fundamentally compatible and therefore to be considered in the same context. In our opinion the most convincing theories of human origins are the ones that integrate the paleontological and genetic approaches. After all, fossils and molecules must ultimately reflect the same thing—the evolutionary history of human adaptation.

In this age of the human genome, some critics of paleoanthropology maintain that so much paleontological data are lost and irretrievable that fossil-based hypotheses are overly prone to fanciful and untestable theories. Zealots of a molecular-only approach also claim that anatomy can provide only a rough idea of ancestral and descendant relationships. To paraphrase a quip by molecular anthropologist Vincent Sarich, they know that their molecules had ancestors while the paleontologists can only hope that their fossils had descendants.

Fossil-only theoreticians, on the other hand, seek refuge in an illusory security of "character states." These are ideal constructs erected for extinct species' anatomy that proponents believe avoid the inconvenient uncertainty of the probabilistic statements of population genetics. Convenient they may be, but the constructs tend to last only until the next new fossil is found. Very much like the ancient Ptolemaic astronomers, the fossil-only theoreticians must make up new rules every time a new discovery is made.

Are we not interested in the details of the evolutionary story that only fossil data sealed at a particular time and place in the past can tell us? And at the same time, is it not a fascinating undertaking to integrate this detail with the undeniable reality of our genetic blueprint, seen both at the species level and in all its local complexity and variability? Our best view of human evolution will be from an integration of all the good data that we can bring to bear on the questions, regardless of the doctrinal turf from which they derive.

Peking Man and Dragon Bone Hill played an important part in transforming paleoanthropology from an arcane paleontological discipline into a modern science that brought in evolutionary biology. The crucible was New York City in the 1940s, and the reaction involved three unlikely protagonists: veteran anatomist Franz Weidenreich, arrived from Beijing in 1941; leading geneticist Theodosius Dobzhansky, who came to Columbia University as professor of genetics from the California Institute of Technology in 1940; and a young physical anthropologist named Sherwood Washburn, still one year away from earning his doctorate at Harvard University when he joined the Columbia anthropology faculty in 1939. In 1937 Dobzhansky had published *Genetics and the Origin of Species*, one of the foundations of the modern synthetic theory of natural selection and an anchor of modern evolutionary biology. In this work Dobzhansky used his and geneticist Thomas Hunt Morgan's experiments and observations on the fruit fly *Drosophila* to extrapolate to other species, including humans. He emphasized *populations* of organisms, not individual "types," in the study of evolutionary biology. Dobzhansky laid out an approach that integrated fossils and genes, and evolutionary biologists have been using it ever since. He wrote: "The problem of evolution may be approached in two different ways. First, the sequence of evolutionary events as they have actually taken place in the past history of various organisms may be traced. Second, the mechanisms that bring about evolutionary changes may be studied."[29] Washburn sought Dobzhansky out early at Columbia, despite the fact that Washburn's former professor at Harvard, Ernest Hooton, was a typologist antagonistic to Dobzhansky's writings on human evolution. Dobzhansky first asked Washburn warily if he had been a Hooton student. Washburn replied, "I do not believe in types and think it is populations which should be compared." Dobzhansky smiled and heartily shook his hand.[30]

When Weidenreich arrived in New York City in 1941, Sherwood Washburn lost no time in getting to know him. He told Weidenreich about his anatomical experiments with rats, in which he investigated the formative effects of muscles on bone. Even though Weidenreich at first asked him, "But what have rats to do with anthropology?" [31] he quickly agreed with Washburn's comparative approach to teasing apart the underlying functional causes of the changes in skull form—changes that he had spent years describing for *Homo erectus*. Discussions with the up-and-coming Washburn may well have influenced Weidenreich's ongoing efforts to understand hominid cranial evolution through comparative studies of dogs, which he incorporated into his 1941 volume on the evolution of the human skull.[32]

Weidenreich and Dobzhansky became aware of each other as scientists on the important issue of race, a topic of overriding concern at the end of World War II as the world tried to make sense of the European Holocaust and horrendous ethnically based genocides worldwide. Weidenreich, who had suffered intense personal and family distress over the course of his lifetime because of ethnic persecution, opined that "physical anthropologists have gotten into a blind alley so far as the definition and the range of individual human races and their history are concerned."[33] He showed that modern populations did not exhibit the preponderance of ideal anatomical traits that propagandists claimed for "pure races," and that mixing between populations had created a "hybrid or multihybrid" character to the human species. Dobzhansky agreed wholeheartedly with these findings and added that genetic determinants should be known for anatomical traits used to assess human evolutionary relationships.[34] Weidenreich integrated genetics into his multiregional model and concluded, somewhat surprisingly, that "not only the living forms of mankind but also the past forms—at least those whose remains have been recovered—must be included in the same species."[35] Yet he steadfastly refused until the end of his days to revise the taxonomy of the hominids that he had done so much to reveal to the scientific community. He cited long-established usage of the old terms, but perhaps he could just not bring himself to abandon the names for which so many of his old friends and colleagues had labored for so long. In frustration, fruit-fly expert Dobzhansky was forced to sink the Javan *Pithecanthropus erectus* and Chinese *Sinanthropus pekinensis* formally into the species *Homo erectus* in 1944. Paleontologist George Gaylord Simpson, another architect of the modern evolutionary synthesis, sneered in 1945 that "perhaps it would be better for the zoological taxonomists to set apart the family Hominidae and to exclude its nomenclature and classification from their studies."[36]

Sherwood Washburn left New York for the University of Chicago in 1947, the year before Franz Weidenreich died. He went on to revolutionize the study of human evolution in America by introducing the "new physical anthropology" in 1951.[37] The central tenets of Washburn's vision were crystallized in New York during the war years of 1939 to 1946 when he integrated the synthetic theory and its genetics from Dobzhansky and then the anatomy of the hominid fossil record from Weidenreich. The Peking Man fossils, though lost, may be said to have played a fundamentally important role in the development of biological anthropology in the United States, far and above their importance in documenting a stage of human evolution. Washburn not only worked hard throughout the rest of his career to ensure that the family Hominidae stayed a part of the research agenda in evolutionary biology, but he also continued to be biological anthropology's most important theoretical link to Franz Weidenreich and the fossils from Longgushan.[38]

CHAPTER 4

The Third Function: A Hypothesis on the Mysterious Skull of Peking Man

Sorting out the "muddle in the middle" requires us to take a fresh look at the cranial anatomy of *Homo erectus*. The most distinctive anatomical differences setting off *Homo erectus* from its ancestors and its descendants are undoubtedly in the skull. Modern humans and our recent ancestors have thin-walled and capacious bony globes that perch atop our spinal columns, holding an enormous, easily injured, semiliquid brain inside. In contrast, the skull that surrounded the *Homo erectus* brain had a massively thick bony wall, enclosing a smaller cranial capacity and exhibiting a low, wide profile. Without the facial skeleton, a skull of *Homo erectus* looks remarkably like a turtle carapace. In fact, field researchers have mistaken fragments of *Homo erectus* skull for turtle shell in fossil excavations. The skull reminds others of a cyclist's helmet—low and streamlined, designed to shield it from blows and to protect the brain, eyes, and ears.

When Dutch anatomist Eugene Dubois first discovered the skullcap that he named *Pithecanthropus erectus* in eastern Java, he was struck with its unusual anatomy. Because there was essentially no fossil record of hominids of any antiquity at the time of his discovery, Dubois and everyone else initially interpreted the skull's anatomy to be indicative of the primitive condition of humankind. As we have seen, Weidenreich interpreted the massive skull of *Homo erectus* as a record of the gigantic ancestry of *Homo sapiens*, believing that a massive skull had to go with a massive body. But as more fossils have been discovered, it is now clear that *Homo erectus* was not a giant; the species just had a very strange skull. And nobody to this day has figured out why.

A Weird Skull and How It Got That Way

Homo erectus skull bone can be technically described as "pachyostotic" (literally "thick-boned"). In understanding how pachyostosis evolved, we can look to comparative anatomy. A few other vertebrate species have or had thick bones, and attempting to understand their adaptations can give us some idea as to the reason for the massiveness of their bones. When comparing animal species that have evolved similar anatomy, we are not looking at traits inherited in common from ancestors, but instead traits that have evolved in parallel for similar reasons. Some species, such as sirenians (marine mammals such as the dugong and Florida manatee) have dense bones throughout their bodies to give them negative buoyancy in water. The ribs of sirenians are essentially ballast. The postcranial bones of *Homo erectus,* however, like ours today are not massively thickened for an aquatic existence.

Among terrestrial animals, extremely thick skull bones are seen in species as diverse as modern bighorn sheep (*Ovis canadensis*) and the Cretaceous dinosaur *Pachycephalosaurus.* The adaptive significance of pachyostosis in these species, based particularly on behavioral observations of male bighorn sheep, is protection of the brain and sense organs during intraspecific competition. Bighorn sheep (and presumably *Pachycephalosaurus* in the

The pronounced cranial thickness of the skull of *Homo erectus.* This is a photograph taken by Davidson Black in the process of preparing and reconstructing Zhoukoudian *Homo erectus* Skull III. In this view the two parietal bones have been removed and a natural endocast of limestone that has a canine of a cave bear imbedded in it just behind the frontal bone can be seen.

past) use the head as an organ of offense, butting heads so forcefully that it sounds like an explosion. Two males run at each other and collide at 30 to 45 miles/hour, generating impact forces of up to 2,700 pounds. The sound can reportedly be heard a mile and a quarter away. What causes such intense conflict among male bighorn sheep? Females. Charles Darwin long ago explained such behavior in species as a result of sexual selection, a type of natural selection that works primarily inside social species and results in same-sex competition for access to opposite-sex mates.

A comparison of the skull bone thickness of male bighorn sheep and the skull bone thickness of *Homo erectus* might at first seem strained. Who would seriously postulate that early hominids might have charged at one another and banged their heads together like rutting sheep? If you asked people on the street (and managed to get them actually to ponder it for a moment), they might suggest that the only modern humans who might be so rash, illogical, and violent as to engage in such behavior for sex would be young adult males. And they would be right. Ample statistics show that 15- to 24-year-old men in the United States die at four times the rate of females from fights, aggressive acts, or accidents that their risky behavior originally made likely.[1] Regardless of their ethnic or socioeconomic backgrounds most disputes between young human males directly or indirectly involve competition for the attention, affections, and affiliation of young females—not all that different from rutting sheep after all. So it is well within the general behavioral capabilities of at least one age group and gender of modern people to fight in a way quite analogous to that of bighorn sheep. We can surmise that it was also possible behavior for *Homo erectus*. Using the head as an offensive weapon may be a less attractive evolutionary option for hominids than for sheep, however. The easily injured human brain is much larger and less well covered by protective bone than is a sheep's brain. We suggest that thick cranial bones in *Homo erectus* might be adapted less for offense than for defense.

Unlike bighorn sheep, human beings have always tended to fight with their hands, and (leaving out gunshots) almost all cases of violent trauma inflicted during nonsexual assaults are to the head. The general pattern of interpersonal attack in humans is to hit the face and head with the hands. Anatomy also suggests that head butting was not a primary adaptation of our ancestors. Pachyostotic species that use the head as a weapon also have cranial outgrowths of bone that evolved for that purpose. Sheep have sharp horns rising out of their thick skull and the *Pachycephalosaurus* had nasty-looking bony knobs projecting around its head like a crown. *Homo erectus* had none of these offensive adaptations. Thus, just from general principles, we interpret the cranial pachyostosis of *Homo erectus* as protective of the brain and sense organs, but defensive in nature. In functional terms we

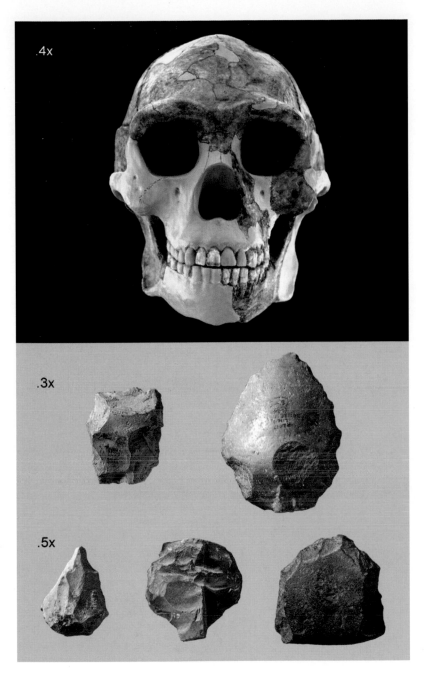

Top: A reconstruction of *Homo erectus* from Longgushan by Ian Tattersall and Gary Sawyer. This view of the skull differs from Weidenreich's earlier anatomical reconstruction in incorporating more robust, probably male, fossils. ***Bottom:*** Stone tools from Locality 1 fashioned by *Homo erectus.* All tools are classified as scrapers except for one chopping tool (first row, right).

View of the western wall of the excavation at Longgushan Locality 1 from Pigeon Hall Cave ("Gezitang"). This cave was first dug by traditional dragon bone diggers and then expanded by later scientific excavation.

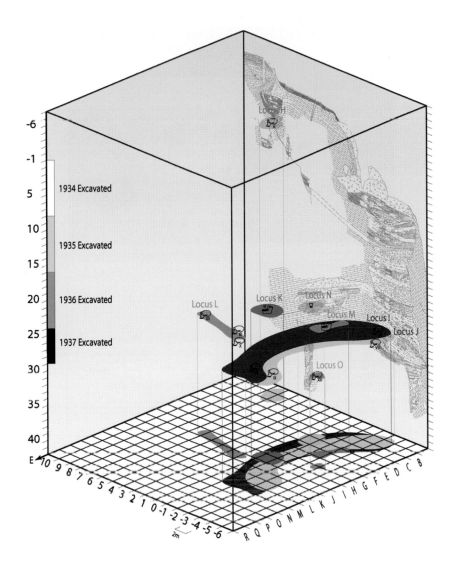

A reconstructed three-dimensional map of the Locality 1 excavation, showing the spatial extent and stratigraphic levels of all "loci" (discovery areas of fossil *Homo erectus* skulls) at the site. Data have been digitized from Lanpo Jia's original plans of the excavation, salvaged by him in 1941. This view is to the southwest of the excavation and shows the stratigraphic profile of the western wall published by a geological team from the Chinese Institute of Vertebrate Paleontology and Paleoanthropology in 1985. When controlled excavation was initiated in 1934, the excavation grid was extended to include earlier loci. The horizontal grid is laid out in 1-meter-by-1-meter squares. Vertical levels were excavated in one meter intervals, shown on the Z-axis by excavation year. The vertical scale has been exaggerated by a factor of two to allow better visualization of the strata and loci. This digitized version of the Zhoukoudian site can be rotated in the computer to appreciate spatial relationships within the site, orient old photographs of the excavations, and guide future research.

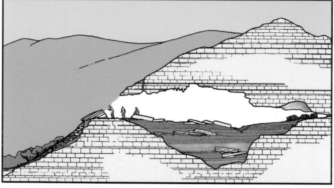

Top: The Pleistocene cave hyena, *Pachycrocuta brevirostris,* was the primary denizen of the cave at Dragon Bone Hill. Its skull is shown here on the right, dwarfing the *Homo erectus* skull to the left, reproduced on the same scale. *Middle:* Abundant stone tools document that *Homo erectus* occupied the site, probably near well-lit cave entrances, in the daylight hours. Fire probably played an important part in hominids' driving the hyenas away from their kills. *Bottom:* As night fell, after the hominids had scavenged what they could, they left the cave to the hyenas.

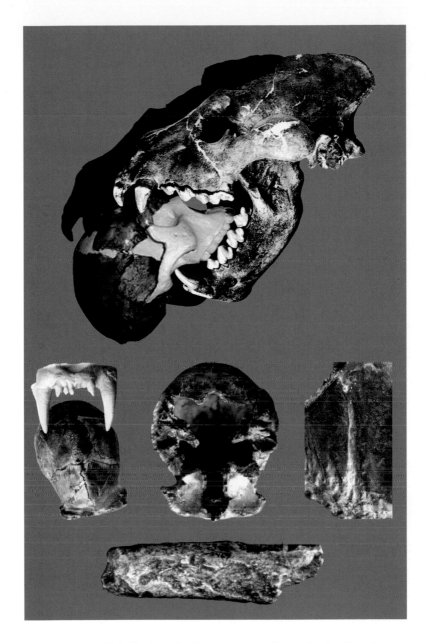

Bite marks on hominid skulls show that hyenas crunched away the face first, then gnawed on the calvaria to crack open the braincase. The shadowy profile represents the second "skull-cracking" bite that exposed the tasty, lipid-rich brain. Some paired puncture marks on *Homo erectus* crania match the canines of other carnivores like this large cat (middle row, left). The enlarged hole at the base of the skull (middle row, middle) was the handiwork of hyenas, not hominid cannibals. A hominid skull fragment shows a hyena bite mark (middle row, right), and a hominid thighbone (bottom row) shows surface damage that matches modern bones regurgitated by African hyenas.

The traditional interpretation of Zhoukoudian—the cave home of Peking Man. We argue that the evidence no longer supports this hypothesis.

The new interpretation—Longgushan as fossil hyena den. Peking Man (camping in the distance) was a fleeting scavenger in the cave but many times entered it unwillingly, as prey.

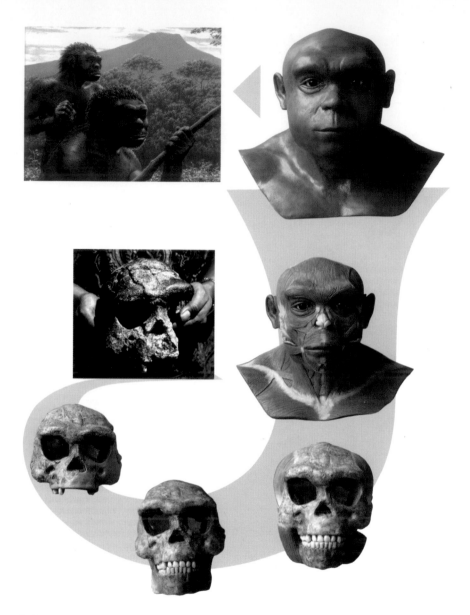

Fleshing out the bones. A reconstruction in stages of *Homo erectus* from Java by Japanese anthropologist Professor Hisao Baba based on the 1.25-million-year-old skull from Sangiran. The final artistic rendering of Javan *Homo erectus* at top left is by artist Jay Matternes. "Java Man" lived in a tropical, riverine, lowland environment surrounded by volcanoes, while "Peking Man" occupied a northern, temperate, karst region with fluctuating climate mediated by the Ice-Age glaciations.

might think of the *Homo erectus* skull as also similar to the defensive carapace of a turtle.

The human head has more small structures, spaces, holes, canals, vessels, muscles, and wiring in its relatively small compass than has any other region of the body. Physical anthropologists and anatomists have been obsessed with describing, explaining, and understanding this complexity for centuries. We now know enough of the human skull's evolution, embryological development, comparative anatomy, and function to feel confident that the major forces affecting human skull form are understood. Thus, when we consider one aspect of the unique skull form of *Homo erectus*—pachyostosis—we must also look at what else this species's skull did for its owners, and how this knowledge helps us understand how it lived.

We interpret evolving hominid skull form as resulting from three major functional imperatives: housing a rapidly increasing brain, serving as the bony anchor for the teeth and the muscles that move them, and, in the case of *Homo erectus*, defending against blunt trauma. All three functions are important to understanding the unusual cranium of *Homo erectus*.

The First Function: Cogito Ergo Sum

G. Elliot Smith, both as an anatomical mentor of Davidson Black and as a describer of the Longgushan skull himself, promoted a view of the primacy of the brain in the evolutionary transformation of the human skull. Paraphrasing French philosopher René Descartes, we might characterize this hypothetical stance of embracing the human brain as the prime mover of hominid skull evolution as "cogito ergo sum"—"I think therefore I am." Less metaphorically, in this theoretical stance natural selection placed a premium on human intelligence, and the brain increased in size and complexity as the human species evolved through time. Anatomical changes that we can observe in hominid fossils record this evolution.

Hominid brain size, as recorded by the space within the skull, increased substantially through time. The earlier three-quarters of our known cerebral expansion were entirely unknown in the first quarter of the twentieth century. This early part of the hominid fossil record has been discovered primarily in Africa and mostly in the latter part of the twentieth century. It shows an increase in cranial volume beginning from a chimpanzee-sized brain in early *Australopithecus*, a fossil hominid genus first published in 1925 but roundly ignored by our protagonists Elliot Smith, Black, and Weidenreich.[2]

The increase in brain size that occurred in *Homo erectus* explains some of the characteristics of its cranial form, particularly the more globular nature

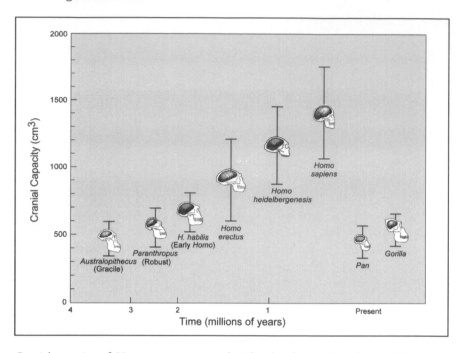

Cranial capacity of *Homo erectus* compared with other hominids and apes. The cranial capacities of early hominids (*Australopithecus* and *Paranthropus*) generally fall within the range of the African great apes (*Pan* and *Gorilla*). When the genus *Homo* evolved, brain size began increasing at a much more dramatic pace. The upper range of cranial capacity in *Homo erectus* overlaps with the lower range documented for *Homo sapiens*.

of the neurocranium (that part of the skull containing the brain) compared to the viscerocranium (that part of the skull containing the teeth). A larger brain had to have a larger skull to contain it. *Homo erectus* has larger, broader, and flatter frontal, temporal, and parietal skull bones than its ancestors. But there are evolved features of other parts of its skull that are not primarily related to brain size increase.

The Second Function: Smaller Chewing Muscles, Teeth, and Faces

The evolutionary transformation of the australopithecines to *Homo* was accomplished by evolutionary change and reduction in the massive teeth and chewing muscles that typified our earliest hominid ancestors. The australopithecines had large teeth, large chewing muscles to move them, and consequently, large bony faces, to serve for anchoring the teeth and attaching the muscles. By the time the australopithecines yielded the evo-

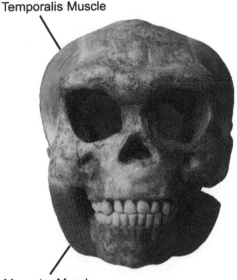

Temporalis Muscle

Masseter Muscle

The *Homo erectus* skull showing the areas of bony attachment of the two largest chewing muscles, masseter and temporalis. Note that the temporalis muscles attach well down the sides of the head from the sagittal keel. Their relative size therefore cannot explain the occurrence of this anatomical feature. The sagittal keel seen in *Homo erectus* is hypothesized here to be related to defensive strengthening of the skull.

lutionary stage to *Homo*, the size of the dentition had diminished. The face had changed significantly. The bone around the nose and extending back to enclose the molars became thinner, and the australopithecine "dish face" disappeared. The bony nose became more prominent, sticking out from the face, the maxillae of which sloped gently backward. The arch of bone holding the front teeth became a smooth parabola, no longer the straight ridge of bone extending from one canine socket to the other as in the australopithecines.

The strong ridges that the australopithecine chewing muscles left on the bones of their skulls decreased in size in the genus *Homo*. The temporalis muscles crept up the sides of the australopithecine skull, sometimes to meet themselves in the midline. When this happened, a vertical ridge of bone known as the sagittal crest formed. It is literally an upright plate of bone to which fibers of the temporalis muscles attach on both sides. Sagittal crests were most common in the robust australopithecines, and occasionally occurred in other, non-robust forms as well. But they are virtually nonexistent in *Homo*, whose small chewing muscles and enormously expanded skull vaults relegate the temporal lines to insignificant linear elevations far down on the sides of the head.

Decrease in the chewing apparatus explains some of the *Homo erectus* skull form—its relatively lightly constructed face, its curved dental arcade in the front, and its generally rounded skull lacking a sagittal crest. But there are still other parts of the *Homo erectus* skull that are not explained by either dental changes or brain expansion. It is to these uniquely *Homo erectus* features that we now turn.

The Third Function: Protection of the Brain, Spinal Cord, and Eyes

Most human anatomists and paleoanthropologists today would agree that the evolution of the brain and the chewing apparatus of hominids is of major importance in explaining the anatomical changes that we see documented in the hominid fossil record. The only problem is that neither of these explanations is sufficient to account for the unique attributes of the strange skull form of *Homo erectus*. We believe that a third function contributed to the evolution of the *Homo erectus* skull, and it is protection.

When people today sustain head injuries they are much more likely to die when their skull has been fractured. What might seem to a casual observer like a relatively minor break can in fact tear blood vessels that tightly adhere to the inside surface of the bone and cause intracranial bleeding. The buildup of blood inside the skull pushes on the brain. Coma and, eventually, death can result.

A common type of fracture seen on modern skulls is an "eggshell" fracture. Concussion by a heavy or strongly wielded blunt object can depress a section of the cranial vault, cracking but not disjointing the bone. The bone springs back to nearly its original shape after the impact—pulled by the attachments of skin, underlying scalp, and muscle coverings—but the damage is done. Branches of the arteries supplying the fibrous coverings of the brain, the meninges, begin to bleed. This blood accumulates as a hematoma in the space between the inside of the skull and the outer meningeal covering, the dura mater. As the hematoma expands and begins to compress the patient's brain, sometimes hours after the injury, neurological symptoms become progressively severe, culminating in loss of consciousness, coma, and death.

In the days before emergency rooms, X rays, and intracranial surgery, people hit hard on the head and suffering from intracranial bleeding survived as best they could. This usually meant not very well. Even if an individual managed to regain consciousness and survive an extradural hematoma, there are frequently residual and significant neurological deficits. Partial paralysis, gait problems, lack of eye-hand coordination, speaking difficulties, or any number of cognitive function disruptions can result. For active Plio-Pleistocene hominids, one could hardly imagine a more debilitating condition, and we might reasonably surmise that traits that would reduce the chances of cranial fracture would provide a significant selective advantage to those individuals possessing them.

Australian researcher Peter Brown has investigated skull thickness in modern and historical Australian aboriginal populations.[3] These peoples show the thickest cranial bones of any members of living *Homo sapiens*. Brown has hypothesized that their thick skull vaults may have evolved by way of their traditional method of settling conflicts.

A man or woman with a bone to pick with another member of the group follows a fixed behavioral code for resolving the conflict, challenging their adversary to a duel with the "nulla-nulla" (a heavy wooden club). Once the bout begins, it continues until one of the combatants wins either by knockout or TKO (his or her adversary is disabled and cannot continue). There are no wins on points or split decisions. Occasionally, entire communities became involved. One ethnographic report of the South Australian Narrinyeri tribal group reported that some one hundred people involved in a general melee were "earnestly endeavoring to knock each other's brains out."[4]

Brown found evidence of healed depressed fractures on the frontal or parietal bones of an amazing 59 percent of female crania and 37 percent of male crania in a sample of 430 Aboriginal crania that he studied. These results mean that depressed eggshell-type fractures occurred in these people and that they survived. But undoubtedly, many others did not. Brown concludes that "behavior of this type must have rigorously selected against those individuals with thinner bones in their cranial vaults and favoured thickened frontal and parietal bones where the blows most frequently occurred."[5]

If Brown is correct and skull thickness evolved in Australians as a result of generations of head bashing, is the model useful for understanding the evolution of pachyostosis and the unique bony excrescences of *Homo erectus*? We believe that it is.

The anatomical aspects of the *Homo erectus* skull that are least explicable in terms of the first functions of brain size increase and chewing apparatus decrease are best explained in terms of the third function—an evolved defense against trauma. We examine each trait in turn.

Experiments that we have undertaken on the strength of modern human bone as a function of thickness have clearly shown that the thicker the bone, the better it is able to withstand forces that would break it. In human biological terms, the thicker a cranial bone, the less likely it will bend in and crack like an eggshell, rending delicate blood vessels and brain tissue underneath. This general protective function of thickened cranial bone is the best explanation for why the *Homo erectus* skull is constructed of bone that is almost twice as thick as that of most modern humans.

The thickness of *Homo erectus* cranial bone is anatomically distinguishable from pathologically thickened modern human cranial bone. Diseases, like malaria that affect the blood and cause the blood-forming bone marrow

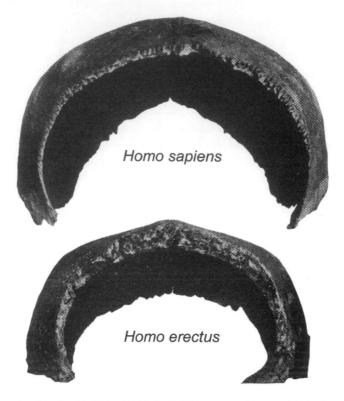

Homo sapiens

Homo erectus

A photograph taken by G. Elliot Smith in Beijing comparing cranial thicknesses of the parietal skull bones, viewed from the front, in modern Chinese *Homo sapiens* (*top*) with Zhoukoudian *Homo erectus* (*bottom*).

to increase in size, can increase bone thickness.[6] The resultant bone is like Swiss cheese, composed mostly of large, marrow-filled spaces, with a very thin skin of compact bone on the outside. *Homo erectus* bone, on the other hand, is like armor-plating. Franz Weidenreich observed that the skull vault bones from Longgushan had thick, solid layers of bone on their inside and outside surfaces. They sandwiched between them the marrow-containing trabecular bone and were, in aggregate, thicker than this softer inside layer.

In a fascinating recent study, Greek anatomist Antonis Bartsiokas investigated the microscopic structure of one of the earliest *Homo sapiens* from Africa, the Omo Kibish I skull from Ethiopia. He found that the thickness of this skull vault was within the range of *Homo erectus* and similarly showed the thickened inner and outer bony armor plating typical of *Homo erectus*, and unlike most *Homo sapiens*. The microscopic structure of the thickened bone also showed a different arrangement—the individual structural elements, the "osteons," were flattened and pressed together. Bartsiokas hypothesized that "perhaps this osteonal morphology is an adaptive means of strengthening the skull against head injuries."[7] So far, *Homo erectus* fossil skull bone has not been similarly investigated, but the possibility exists that even its microscopic structure will make sense in terms of a defensive adaptation to protect against blunt head trauma.

In addition to the general thickness of its bones, the *Homo erectus* skull also has a number of unique bony structures. Franz Weidenreich gave these Latin names—*torus supraorbitalis, torus angularis, torus occipitalis,* and *crista sagittalis*—better referred to as the sagittal keel. The first three bony thickenings form a ring starting above the eyes, extending back around the head above the ears, and meeting on the back of the head. The sagittal keel forms a thickened bony mound from the middle of the forehead extending back over the crown of the head to meet the horizontal ring of bone in the back. The characteristic flattening of the *Homo erectus* skull falling away to both sides of the sagittal keel provides additional strength to the skull vault.

A forensic review of man's inhumanity to man provides ample evidence of how important these bony adaptations would have been to a hominid routinely subjected to blunt trauma to the skull. An American surgeon named E. R. LeCount classified the types of fractures that occur when people are hit hard on the head.[8] A heavy blow falling directly on the top of the head tends to cave in the bone protecting the channel of venous blood draining along the midline of the brain known as the superior sagittal sinus. If this structure is damaged there is bleeding into the space between the brain and its outer covering, the dura mater. A so-called subdural hematoma (blood collecting under the dura mater) can compress the brain, causing defects of function, and it is potentially fatal. LeCount hypothesized that the strongly constructed midline of the human skull is adapted to protect against this type of damage. In most *Homo erectus* this adaptation appears in exaggerated form as the sagittal keel, a low, rounded thickening of bone running from the front of the skull to the back.

However, blows do not usually rain down from above on the heads of antagonists in a physical disagreement, but are instead delivered more or less at eye level. The battered skulls of Bosnian and Croatian victims of genocide, for example, uniformly show damage to the regions around the eyes, on the sides of the head, behind the ears, and at the back of the head.[9] This pattern of damage is exactly the location of the ring of tori as seen in the *Homo erectus* skull. LeCount saw the same areas in the modern human skull as protection against the most common injuries resulting from blunt trauma to the head.

Another surgeon, René Le Fort of France, studied the pattern of facial fractures in modern people, and his conclusions are also instructive.[10] Le Fort classified the types of fractures that he observed. A Le Fort Type 1 fracture is one that results from a blow to the upper face that breaks the bone surrounding the eye socket. Direct blows to the brow frequently break the orbit, the bone forming the roof of the bony space that holds the eyeball and its muscles. *Homo erectus* has a remarkably straight roof to the orbit, an anatomical peculiarity that until now has defied a functional explanation. Extending back to the strong base of the skull from the heavy supraorbital tori this trait would have helped *Homo erectus* individuals avoid Le Fort Type 1 breaks. Weidenreich himself, in his posthumous publication on the Javan Ngandong crania,[11] suggested that the massively projecting ridges of bone above the eyes of *Homo erectus* likely had a protective function. Most recently, John Hawks of the University of Wisconsin used three-dimensional simulations of impact trauma to the head and concluded that having a larger supraorbital torus significantly enhanced protection of the upper face and eyes.[12]

The straight and relatively unangled face to neurocranium hafting that is characteristic of *Homo erectus* would have prevented Le Fort Type II and III fractures, very debilitating breaks that result from separation of the facial skeleton from the braincase. A strong blow to the reinforced *Homo erectus* face would have resulted in soft tissue damage, including perhaps less serious fracture of anterior parts of the maxilla, but fractures mobilizing the teeth or the zygomatic arches would have been reduced. Anterior blows to the face would also have resulted in fracturing incisors particularly, and these teeth in *Homo erectus* also show a reinforcing thickening of enamel on their lingual sides that could have prevented loss of dental function. *Homo erectus* shows a high incidence of such "shovel-shaped incisors."

An old boxing adage warns away those potential pugilists with a "glass jaw." Indeed, fracture of the mandible is a serious injury in those on the receiving end of a barroom punch to the chin or lower face. A broken jaw renders active chewing painful and difficult, if not impossible, and today it requires surgery and wiring together of the broken sections of bone. Clearly a jaw fracture would have been a life-threatening event for a *Homo erectus* individual, who without surgery would not have been able to eat solid food. We suspect that a large number of *Homo erectus* unfortunately died this way. The anatomy of the *Homo erectus* mandible shows a unique thickening of bone on the side of the jaw exactly where it most commonly breaks from trauma. Weidenreich in his monograph on the Longgushan mandibles named this thickening of bone on the inside of the mandible the *torus mandibularis*. At first thought to be pathological, the *torus mandibularis* makes most anatomical sense as another defense against trauma to the lower face.

A point of *Homo erectus* anatomy that is particularly convincing to us regarding the protective pachyostosis hypothesis, concerns the course of an artery that can be traced on the inside of the skull. The middle meningeal artery is a branch off the maxillary artery and runs up inside our temple. Its vulnerability to damage is the reason that the baseball batting helmets have a little protective flange extending down on the side of the head facing the pitcher. The bone overlying the middle meningeal artery at a region of intersecting sutures known as the pterion is particularly thin in modern humans. It is partially protected by the overlying chewing muscle, the temporalis, but a good shot in the temple is likely to break the bone here and tear the middle meningeal artery. An arterial tear is even more dangerous than damage to a venous sinus because the blood in an artery is under higher pressure and can bleed out more rapidly. Usually damage at the pterion results in a large amount of blood pooling on the outside of the dura mater (an extradural hematoma) and rapid loss of consciousness or coma.

The *Homo erectus* skull is not particularly thick at the pterion, and if this region is susceptible to damage in modern humans, we would expect it to have been even more so in *Homo erectus*. Observations by Franz Weidenreich on the unusual anatomy of the middle meningeal arteries in *Homo erectus*, until now unexplained, may provide the answer.

In modern humans the middle meningeal artery divides into a large branch that runs forward on the inside of the skull under the pterion, and a smaller posterior branch that runs backward. But in *Homo erectus*, the anterior branch is miniscule compared to the quite large posterior branch. Weidenreich was so struck by this anomaly in *Homo erectus* that he devoted an entire paper to it.[13] We think that this anatomical trait of *Homo erectus* is a result of natural selection to withstand the effects of breakage in this area of the skull. If for developmental and structural reasons (because perhaps it is a convergence point of skull sutures) the region of pterion could not easily be thickened during evolution—especially as the cranial vault was expanding with a larger brain—it makes sense that *Homo erectus* adapted to minimizing bleeding in this area should this artery be torn. By redirecting blood flow to a posterior course, under stronger skull vault bone, the chances of incurring a fatal and debilitating epidural hematoma would have been substantially reduced. This region of the temporal bone also underlies the large temporalis chewing muscle that can partially cushion a blow to the head at this point.

The back of the skull of *Homo erectus* suggests to us that blows from behind were a major factor in human evolution. The *torus occipitalis* overlies and protects the confluence of venous sinuses inside the back of the skull, branches of the posterior cerebral artery supplying the brain, the occipital lobe of the cerebrum, and the cerebellum. Damage here can result in blindness if the occipital lobe is affected, or inability to walk, stand, or move in a coordinated fashion if the cerebellum is injured. The *torus angularis* overlies and protects the sigmoid venous sinus inside the cranial cavity as it conducts blood into the internal jugular vein exiting the base of the skull. It also helps to protect the ear region from behind.

Blunt trauma to the back of the head is a common cause of death today. The pattern of injury demonstrates exactly what we believe *Homo erectus* anatomy evolved to prevent.

Paleopathological Evidence in Support of a Defensive Function for *Homo erectus* Skull Form

Franz Weidenreich was a trained medical doctor and had worked most of his career in medical institutions in Germany. It is likely that he had more than a passing familiarity with the effects of head trauma, although he did not publish on this subject early in his career. We believe that Weidenreich's identifications of healed depressed fractures on the skulls of *Homo erectus* should be taken much more seriously than they have been. Consequently, we undertook a systematic reexamination of Weidenreich's evidence for cranial bashing in *Homo erectus* using all the casts of the excavated remains from Longgushan.

In Weidenreich's final analysis, he attributed some ten depressions or defects in the skulls from Longgushan to hominid agency.[14] He recanted on some earlier claims, ascribing this damage instead to carnivores. Other damage is clearly geological—crushing from the weight of overlying sediment and impressions in the bone from rocks pushed into the fossilizing bone by the enclosing sediment. But a number of the remaining depressions in the *Homo erectus* skulls from Longgushan match closely the size, form, and even location of healed depressed fractures seen in modern human skulls.

The face and lower jaw frequently bear the brunt of frontal assaults. Are there any anatomical indications that *Homo erectus* evolved to taking it on the chin, as would be expected if our argument is correct? Broken jaws in modern barroom brawls frequently occur just behind the chin region.[15] In *Homo erectus* this area of the mandible on both sides of the

---

Facing page
Healed depressed fractures are depressions in the skull resulting from a blow heavy enough to break the outer table of bone but not to cave in and displace a fragment of bone. The top view is a contemporary *Homo sapiens* (United States) showing a healed depressed fracture near the crown of the head along the sagittal suture. *Middle* and *bottom* views show the depressed fracture in *Homo erectus* Skull X from Longgushan in a similar location.

jaw is strengthened to form a thickened mass of bone termed by Weidenreich the *torus mandibularis*. Like the various cranial tori there has never
been an adequate functional anatomical explanation for the *torus
mandibularis*. It does not serve for the attachment of any muscle and its
thickness is not in the area of support for chewing strength. We believe
that the *torus mandibularis* is also an adaptation to withstand trauma to
the jaw and lower face.

Broadly speaking, cranial pachyostosis in *Homo erectus* evolved as a result of sexual selection, a subset of principles of Darwinian natural selection that comes into play in social species, the sexes of which compete for
mates. *Homo erectus* uniqueness in skull form then represents a detour in
the broad march of human evolution along the course of an enlarging
brain and a decreasing emphasis on large teeth.

Is There a Fourth Function?—Cooling the Enlarged Hominid Brain

If our hypothesis on the meaning of *Homo erectus* skull anatomy is correct,
and *Homo erectus* as a species was ancestral to later *Homo sapiens*, why did
thickened cranial bones evolve out of our biology? If a thickened skull was
adaptive for *Homo erectus* when these hominids got hit on the head, why
did evolution discard it for us? If modern children had thicker skulls, significantly smaller numbers of them would suffer serious head injuries when
they crashed on their skateboards, bicycles, and snowboards, for example.

As we saw in hypotheses on the appearance of pachyostosis in *Homo
erectus*, the first two functions of cranial evolution—increased size of the
brain and decreased size of the chewing apparatus—do not help explain its
origins. Neither do they help explain its disappearance. A counteracting
selection force, or a combination of such forces, is important to identify in
attempting to understand how and why cranial pachyostosis disappeared
in more advanced hominids.

Anthropologist Dean Falk has hypothesized that the heat generated by
the extremely enlarged human brain became a significant physiological
factor in evolution.[16] In her "radiator brain hypothesis" she proposed that
the pattern of venous blood drainage in the head became reorganized to
cool the brain. Many small holes known as emissary foramina pierce the
skull and serve for the passage of veins from the surface skin to the large
venous sinuses inside the skull. Blood cooled by heat exchange from evapo-
rating sweat on the scalp moves into the venous sinuses. Falk discovered
that emissary foramina are much more common in large-brained *Homo*
species than in small-brained australopithecines. The deduction then is

that cool scalp blood flows back through the skull bone, where it cools the brain and keeps it at optimum temperature. Falk's hypothesis is still under debate, but it does explain some important aspects of hominid cranial anatomy in the evolution of a large brain.

We suggest that the radiator brain hypothesis may also explain why skull thickness in *Homo erectus* decreased as this species evolved. A thick skull would have been substantially more difficult for low-pressure and delicate emissary veins to pierce, thus making it more difficult for the enlarging brain to be cooled adequately. Natural selection may well have favored a thinner skull for this reason as the brain increased in size and metabolic heat output.

Our explanation for the thick skull of *Homo erectus* is a hypothesis of exclusion—it simply makes the most sense of any possible reason we can think of. But the behavioral implications of the hypothesis will be disturbing to many who may want to believe than humanity has a basic adaptation for cooperativity and sociality. We agree, but our behavioral evolution was more complex than can be summarized in one or two words. *Homo erectus* still has much to teach us about the evolution of our behavior, and research continues.

In the next chapter we turn to the Longgushan Cave's primary evidence of behavioral complexity in *Homo erectus*—the use of stone tools and, above all, fire—whose effective use may have been the driving force behind the brain's remarkable evolution.

CHAPTER 5

The Adaptive Behavior of the Not-Quite-Human

A quartz stone tool found by Gunnar Andersson in 1921 was the first clue that fossil hominids would be discovered at Dragon Bone Hill. But during the early years of the excavations, carried out under the supervision of paleontologists Otto Zdansky and Birger Bohlin, no archaeological remains were reported. In 1930 Davidson Black wrote that "though thousands of cubic meters of material from this deposit have been examined, no artifacts of any nature or any trace of fire" were discovered.[1] Was this because they were absent in the interior reaches of the cave, where the first fossils were mined, or was it because these small irregularly broken pieces of quartz just went unrecognized amid the massive rubble resulting from the quarrying for bone? We determined to investigate this question because some important deductions concerning the behavior of *Homo erectus* depend on its answer.

The Excavated Evidence

The first methods used to extricate the fossils from Longgushan were dynamiting the sediments to blast the fossils loose, quarrying the debris by pickax, removing adhering sediment with hammers, chisels, and metal probes, and finally sieving the debris for small fossils that might have escaped detection. Pei and Zhang report that in the excavations between 1927 and 1928 "stone artifacts and the materials of utilized fire were not researched."[2] That the blasting was less than controlled is suggested by the fact that the entire Temple to the Hill God on Longgushan was accidentally blown up during one of the early field seasons.[3] There was no map-

The early years of research at Longgushan saw the use of setting black powder charges and blowing them up in order to break up the sediment. Mapping of fossil and artifactual specimens recovered from the site during these years can only be approximated. This photograph shows workers drilling a hole for setting explosive charges in 1927.

ping of the fossils' location except to indicate their stratigraphic location on a geological cross section of the cave and to designate their horizontal location by a loosely defined "locus." The sediments were sieved and picked over in woven baskets to find all fragments of fossil bone, but stone artifacts may have been discarded along with the unidentifiable bone fragments. It was not until the 1934 field season at the cave site that excavators painted a three-dimensional grid in units of one meter on the horizontal sides of the cave and one meter on the vertical. The grid was painted onto the rock, and fossils and stone artifacts were recorded and mapped in a controlled manner. Despite this method of collecting the data, in all the years since 1937—the last year of the prewar excavations—a comprehensive map of the excavation had never been compiled. This was a major goal of our research at Longgushan.

Our colleagues at the Institute of Paleoanthropology and Paleontology in Beijing, Dr. Qinqi Xu and Mr. Jinyi Liu, undertook with us a dedicated search of the institute archives for the catalog of these excavations. All that seem to have survived are typewritten summaries originally transcribed by Lanpo Jia in 1941. He was allowed by the occupying Japanese army to

Excavation grid marked out with white paint near the end of the field season, November 1935. View is toward the north, where a plank walkway extends over the entrance into the Lower Cave. Squares measure one meter by one meter. The lettered squares are the "0" line where squares are labeled (A,0), (B,0), etc. Rows of squares to the south increase in number, e.g. (A,1), (A,2), etc., and rows of squares to the north decrease in number, e.g. (A,–1), (A,–2), etc.

continue working in the Cenozoic Research Laboratory at the Peking Union Medical School after Weidenreich left. During the day he managed to transcribe his own notes and maps onto toilet tissue and then smuggle them out past the guards. All the other original records seem to have been lost following the closing of the Peking Union Medical College during the war.

Before he died in 2001, Lanpo Jia allowed Jinyi Liu to photocopy his notes and excavation maps of Zhoukoudian for the purpose of constructing an overall map of the Zhoukoudian excavations for the first time. We reviewed all the published reports of the excavations as well as many unpublished photographs in the American Museum of Natural History Library, where they have been kept since Weidenreich's death in 1948. We used these data to construct a composite map of the site. With it we have been able to locate all the levels of the excavation shown in the many surviving photographs of the site as well as plot the positions of all 15 loci at which hominid fossils were discovered.

Archaeologists use the spatial patterning of artifacts and larger features of a site to interpret past cultural behavior. In the case of Longgushan, both the early methods of excavation and the later loss of much of the data

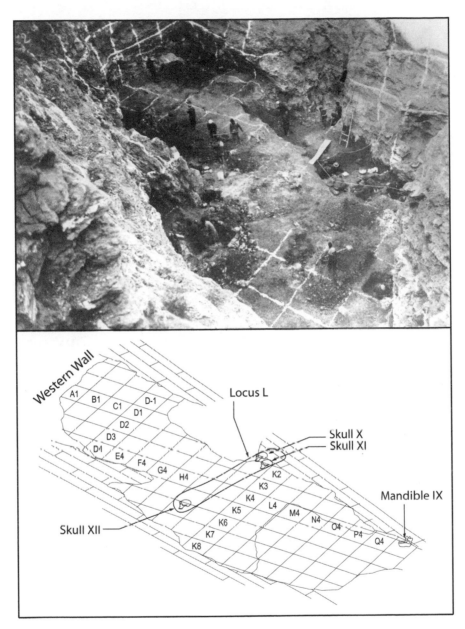

Discovery of Locus L at Dragon Bone Hill, viewed from the southwest. *Top:* Excavation of Skulls X, XI, and XII in Locus L, Level 25, Layer 8/9 of Locality 1. Area enclosed by rope shows excavation of Skulls X and XI and area in left foreground is where Skull XII was found. (Photograph taken November 15, 1936.) *Bottom:* Excavation plan view of Locus L showing location of the three hominid skulls discovered in 1936 within Locus L, as well as Adult Mandible IX, discovered in 1959, at Level 27 of Layer 10.

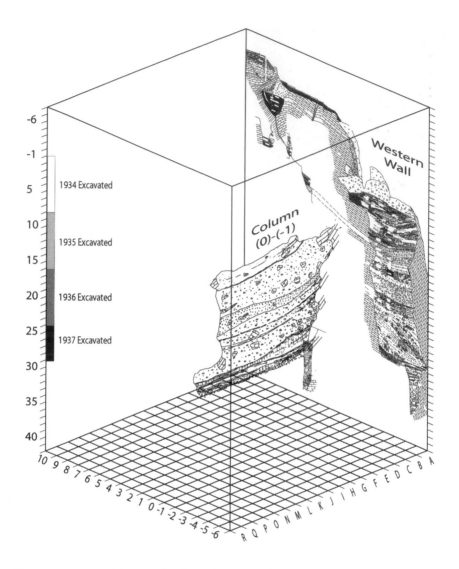

Computer-generated image of a three-dimensional plan of Locality 1 from data preserved by Lanpo Jia, showing locations of major stratigraphic columns and the grid system used in the excavation. A colored version of this diagram complete with many loci and major hominid finds appears in the color insert. The vertical scale is doubled to allow visualization of stratigraphic detail.

from the excavations preclude detailed interpretations of many aspects of the site. For example, did *Homo erectus* leave their stone tools near the entrance of the cave, where we could presume that there was light from the outside, or did they penetrate more deeply into the cave, displacing the resident hyenas, perhaps with the use of fire? Our data are just not sufficient to answer this question.

We can make some reasonable deductions about general aspects of *Homo erectus* archaeology at Longgushan from what data have survived. The artifacts are found at all levels of the site, from the lower parts to the uppermost. The artifacts seem to cluster with the deposits of burned bones at the site, implying that fire and stone tools are associated. And, perhaps most importantly, numerous stone tool cut marks are to be seen on the fossil bones from Longgushan, allowing us to make the connection between the stone tools and the functions to which they were put by *Homo erectus*.

The Types of Tools and the Raw Materials

Caves formed by percolating groundwater in limestone, referred to as "karst" (from a Serbo-Croatian word describing such areas along the Dalmatian coast), are very poor sources for the sort of crystalline rocks that make good stone artifacts. The hominids at Longgushan thus had to bring raw materials for tools in from afar. Many seem to have come from the river gravels of the Ba'er (or Zhoukou) River. Others were apparently picked up by hominids walking farther afield.

Wenzhong Pei, the veteran researcher, and a colleague, Senshui Zhang, in 1985 described some 17,000 stone artifacts from the Longgushan excavations. They reported that some 44 different types of stone were used as raw materials, but by far the largest number (89 percent) were made from quartz. Zdansky had seen broken shards of quartz in his quarrying but he had thrown them out, considering them pieces naturally eroding from the quartz veins in the cave walls. Pei collected isolated flakes of quartz in 1929 and 1930, but it was not until 1931 that an abundance of stone tools was recognized and named the "Quartz Horizon 2" in the eastern part of Longgushan. These artifacts were found in association with hominid fossils in Locus G, and with ash-like layers of sediments thought at the time to be remnants of hominid fire.[4]

The first widely recognized stone artifacts at Longgushan were found in the part of the deposit that is termed "Gezitang" or Pigeon Hall. It is the east-facing artificial opening made by early bone miners and is close to what is thought to have been the original opening of the Longgushan cave when *Homo erectus* was there. The 1931 discovery by Pei of many clear and

The most common stone tools at Longgushan Locality 1 are small, sharp flakes of quartz, several of which are shown here. These tools were used to cut meat off bone and for other purposes requiring finer cutting than chopping tools allow.

unambiguous artifacts was taken by some researchers to indicate that there was a higher density of artifacts near the old entrance of the cave than in the areas farther back from the cave mouth. While such an idea seems to make sense, because hominids would have had natural light near the front of the cave for their implement-related activities, it is difficult to demonstrate. Excavation techniques prior to 1934 had not focused on recovery of artifactual remains, and surviving excavation data are just too sparse to show this type of distribution of artifacts. What reliable data have survived show that artifacts are quite uniformly spread throughout the vertical extent of the Longgushan deposit.

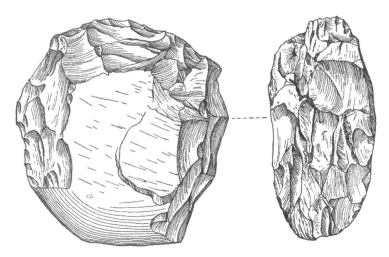

A large chopping tool made from stream-rounded quartz cobble discovered at Longgushan Locality 1. Such heavy stone tools were likely used by *Homo erectus* to dismember animal carcasses.

In 1932 Pei began a collaboration on the archaeology of Longgushan with Pierre Teilhard de Chardin. Teilhard was a widely respected, globe-trotting Jesuit geologist and prehistorian who was allowed by the Catholic Church to pursue research for some 20 years in China. He had also been in England for the Piltdown Man "discoveries," and before that had worked in the famous archaeological caves of southern France. He now worked for the Geological Survey of China in the capacity of a consultant. Pei and Teilhard examined all the archaeological finds up to that point from the cave and concluded that there were three "cultural zones" preserved at Long-gushan.[5] The oldest and most primitive was "Zone C," found in the lowest parts of the excavation. "Zone B" was above that in the excavation, par-ticularly at Locus H, and was typified by hominids' use of better raw ma-terials, such as chert (flint), for the artifacts, and by better techniques of chipping and forming the tools. The highest level was poorly represented in the excavation but was presumed to be the most advanced. This work estab-lished the basis for the excavations, now with a much more archaeological bent, that Pei carried out between 1933 and 1938 at Longgushan Localities 4, 13, 15, and the Upper Cave.

In an overview of the archaeology, Teilhard wrote in 1941 that the "rich lithic industry of Choukoutien" could be divided into two parts: a much larger part of small, quartz flakes, which he described as "splinters of crushed vein-quartz pebbles"; and a less numerous component made up of larger tools, many made from the relatively poor raw material of sandstone— "entire pebbles or boulders, retouched in an elementary way."[6] We might think of these as the ancestors of the penknife, scalpel, and paring knife on the one hand, and the machete, ax, and cleaver on the other. Interestingly, the latter component did not contain any of the tear-shaped, bifacially flaked hand axes that were known from the earliest archaeological levels of Europe. And the assemblage of tools certainly did not contain anything like the sophisticated scrapers, awls, burins, and blades that European ar-chaeologists had discovered in association with early *Homo sapiens*. Teilhard concluded, "in contrast with the already 'steaming' West, early Pleistocene Asia seems to have represented (on account of its marginal geographical position) a quiet and conservative corner amidst the fast evolving human world."[7] We will return to the important behavioral implications of two components of the stone tools below.

Tools of Stone? Tools of Bone?

Teilhard's travels around the world and voluminous correspondence with colleagues ensured that the leading researchers learned about new discov-eries in China by firsthand accounts. One of Teilhard's oldest friends and colleagues in France, the eminent archaeologist Henri Breuil, learned of Pei's and Teilhard's discoveries by letter. Breuil was fascinated. In the fall of 1931 Davidson Black arranged an invitation for Breuil to come to China to pass judgment on the stone artifacts that had been excavated so far. Breuil concluded that "the artificial [i.e., cultural] nature of these specimens is already evident."[8] But what kindled Breuil's imagination more than the stone artifacts, which to his European eye seemed extremely primitive, was the possibility that this early form of hominid had used bone tools as its primary manner of cultural adaptation.[9] A base of a fossil antler from Longgushan that Teilhard had brought with him to Paris in 1930 had first given Breuil the idea. The antler he "recognised immediately as burnt and made into a tool by blows from stone implements."[10] During his 1931 trip to China he saw many more examples of what he considered bone and antler tools made by Peking Man.

Breuil's original idea did not sit well with a number of the other researchers at Longgushan. Still, his ideas were well argued and his reputation alone required that he be given an audience. That some people were listening is shown by Breuil's receiving a formal invitation in 1934 to return to China to collaborate with Pei on a detailed study of the bone tools. That collaboration did not end with Pei as a coauthor on the ensuing publication because he failed to be convinced by Breuil's arguments. The 1939 monograph on the presumed bone tools from Longgushan was authored solely by Breuil.

Breuil's hypothesis on the bone archaeological industry at Longgushan has been largely forgotten. He hypothesized that Peking Man had used the mandibles and isolated teeth of carnivores as weapons, asking rhetorically, "what [could be] more natural than to try to steal their arms and turn them against their owners?"[11] He ascribed the breakage patterns of many fragmentary fossil bones and teeth to the actions of hominids. We now know from comparative studies of modern hyena dens that many of these types of breaks can be, and at Longgushan probably were, made by animals. The many isolated and pointed bone fragments that Breuil thought were produced by hominids' flaking of bone, for example, are identical to the types of bone refuse produced by hyenas today. And Brueil incorrectly identified a fossil rhinoceros upper foreleg bone (the humerus) with multiple, raking carnivore bite marks, as indicated by their clear U-shaped cross sections, as a *Homo erectus* "cutting table." But Breuil's work remains an excellent source of information on the now scattered and lost bones from the site, and some of his valid observations have been unjustifiably thrown out.

Breuil was the first researcher to notice the telltale signs of stone-tool cut marks on bone. One of Breuil's photographs of a fossil antelope foot bone

Archaeologist Henri Breuil (*center*) became involved with the Dragon Bone Hill research through Teilhard de Chardin (*left*). They are shown with a third, unidentified, man at the site on May 4, 1935. Although Breuil developed the now discarded hypothesis of extensive bone-tool use by the Zhoukoudian hominids, his influence also led to the controlled excavation and mapping of the site.

showed "fine cuts made by a stone tool."[12] He further noted that the bone had not been gnawed.

The importance of Breuil's pioneering observations was that they had the potential of connecting stone tools, which were indubitably made by hominids, with animal bones. Not only did this establish a direct ecological link between *Homo erectus* and a species of animal, but how the cut marks were oriented and where they were on the bone could tell a lot about what the hominid had been doing with the tool and what he or she was trying to obtain. The bone evidence showed not what Breuil had originally emphasized—that the bones themselves were tools—but instead the *effects* of stone-tool use, and therefore one of the most important components of the past that any paleoanthropologist wants to discover—behavior.

Archaeologist Lewis Binford and colleagues reexamined a number of the fossil bones from Longgushan that had survived the early excavations and some that had been newly excavated by the Chinese. He had the advantage of much experience in discriminating stone-tool cut marks from animal bite marks on bone from archaeological sites. He saw a large number of cut marks, some of them overlying bite marks made by carnivores (thus implying scavenging by *Homo erectus*). Hominids had used small sharp stone flakes to cut off pieces of meat from haunches of animal carcasses and to remove the tongues from animal heads. Binford never observed any carnivore bite marks overlying the stone tool cut marks of hominids, implying that the hominids had not run down their meat themselves but instead had scavenged the prizes of carnivores.[13]

Brueil also observed a number of bones that had been battered, not cut, with stone tools. Binford's observations paralleled this discovery as well. He noted that there seemed to be two patterns of bone modification by hominids: cut bone and battered bone. The two constellations of hominid-modified bone at Longgushan matched the two overall patterns of stone tools noted first by Teilhard: small, sharp flake tools used for cutting, and large, dull tools used for hacking and smashing. Stone tools at Longgushan were apparently used to cut or hack up animal carcasses before eating, or otherwise using, parts of them.

Evidence of Fire

Fire can occur naturally, as when lightning strikes ignite dry grasslands (still a frequent occurrence in sub-Saharan Africa), or it can be intentionally set and controlled by humans. Fire, more than any other cultural attribute, has been considered the hallmark of humanity. But *Homo erectus* seems to have had a relationship with fire that was unique—they were partially in control of its power, unlike any other animal species, but still in awe of it, unlike modern humans.

Dragon Bone Hill is one of the earliest sites documenting the use of fire by hominids. Davidson Black reported the first discovery of evidence of fire the same year that the first undisputed stone tools were found—1931,[14] but Henri Breuil and Teilhard de Chardin were probably the first to make the original observations.[15] Black adduced four lines of evidence in support of his argument that *Homo erectus* had used fire: carbon deposits, ash accumulations in hearths, burned bone, and fire-cracked stones (presumably used around campfires). All of these aspects of the evidence for fire at Longgushan have recently been reinvestigated by multidisciplinary teams. We look at each of the types of evidence in turn.

In the Lower Cave deposits, a blackened level of sediment was found by excavators. Black took some of the sediment to a chemist at Beijing University, who analyzed it and confirmed it to be carbon. This was an important result because a number of elements, including manganese and iron, can stain sediments and give them a black color. Black made the deduction that the carbon residue was derived from the charcoal left over from a hominid campfire.

A new study has proved Black wrong in his deduction about the carbon level. Paul Goldberg, a geologist at Boston University, and his colleagues reanalyzed the black sediment in the remaining western wall of the cave excavation.[16] They determined that the sediment is indeed carbon, but interestingly it comes from a low stratum of the cave (Level 10) in which flowing and standing water were the main cause of sediment deposition— not a place where any early hominids would be building a fire. The carbon level was one of many finely laminated strata laid down by standing water. Further analysis showed that the carbon was organic residue of undecayed plant remains covered by water, and not the remains of charcoal at all.

Sediments in the excavation that had originally been identified by Teilhard and Pei as light-colored silts were reinterpreted as ashes after 1931 when it was determined that fire had been present at the site. Again, it is likely that Breuil's extensive experience in French Paleolithic cave excavations, in which he routinely found large accumulations of wood ash from early human hearths, influenced this reanalysis.

In 1998 geochemist Steve Weiner of the Weizmann Institute of Science and his colleagues took another look at these presumed ashes from Longgushan.[17] For comparison they used their in-depth analyses of the undisputed Neandertal hearths at Hayonim Cave, Israel. They found that the Chinese sediments lacked the telltale phytoliths that are so common at Hayonim. Phytoliths are small bits of calcite in the tissues of many plants, which provide support for leaves and stems. When fires have been stoked with tons of wood over many years, there is a substantial residue of their phytoliths in the ashes. Weiner and his colleagues discovered that the light-colored sediments from Dragon Bone Hill contained no phytoliths. They therefore could not be ashes from wood fires fed by hominids. Further analysis showed that the sediments were in fact fine windblown sediments known as "loess" that had been reworked and deposited by water. The original identification of the sediments as "silt" by Teilhard and Pei had not been so off the mark after all. But, most significantly, it was obvious that the evidence for *Homo erectus* hearths had gone up in smoke.

The evidence for fire that has best withstood scrutiny has been burned bone. From an anthropological standpoint this evidence is doubly important because it establishes not only fire's presence in the cave but also clearly indicates its use. Lewis Binford and Nancy Stone noticed fire-cracked upper teeth of a horse in the Longgushan collections and deduced that *Homo erectus* had roasted horse heads. Other isolated bones of many species of large mammals showed evidence of having been burned while fresh, strongly implying cooking and eating by hominids.

A subset of burned bones from the site was colored blue, turquoise, or slate gray. Weiner and his colleagues undertook some experiments and discovered that only *fossil* bones heated to 600°C turned color like this (fresh bones either blackened or turned to ash). The conclusions from this line of evidence were that fires had burned at Zhoukoudian and that fossil bones had been exposed on the ground at the time. Rare fire-cracked stones, too large to have been washed into the cave, supported this evidence. Natural fire in a cave that still had a fair amount of standing water in it, even if there was a large accumulation of potentially combustible bird and bat guano, is much less likely in our opinion than the probability that hominids introduced fire into the cave. The geological context of burned bone, the evidence that some bone was burned while fresh—likely as incidental to roasting meat—and the presence of fire-cracked stones all argue that hominids used fire at Longgushan.

Implications for Behavior

What do the various lines of evidence mean in terms of what *Homo erectus* actually did at Zhoukoudian? Some archaeologists have suggested that the apparent sameness of the stone artifacts from the lowest levels of the cave site to those at the top show that *Homo erectus* was a very slow-witted species. Others have pointed out that the hand ax, a tear-shaped stone tool with a characteristic double-faced cutting edge, never made it to China, even though it was characteristic of much earlier *Homo erectus* sites in Africa. Was *Homo erectus* at Longgushan particularly slow on the uptake, or are there other explanations?

The raw material for stone tools at Longgushan may provide part of the answer. Quartz is an abundant crystalline rock that gives a sharp edge when broken, but it is a notoriously poor stone for flaking into large or complex tools. Quartz is shot through with cleavage planes that cause the stone to break into unpredictable shapes, frustrating even the most adept or artistically ambitious of stone-tool knappers. The Longgushan hominids had to settle for small flakes of quartz, which they used for slicing and for scraping muscle off bone. For bigger jobs, such as cutting through the rib cage of a deer carcass, they used a hefty chopping tool made of sandstone. Sandstone occurs in the Zhoukoudian region, but it is not a crystalline rock—it does not fracture like thick glass or give a sharp edge. But for cracking ribs by using brute force, it is effective. Sandstone just cannot be made into a recognizable bifacial hand ax. Thus, we do not think that too much should be made of the perceived deficiencies of the Longgushan stone tools because of the natural geological limitations of Dragon Bone Hill. We believe that the archaeology of Longgushan reflects the general capabilities of the species *Homo erectus*, modified to fit local conditions.

There are other explanations for the rarity of bifacial tools in the Pleistocene Epoch in China. Paleoanthropologist Geoffrey Pope proposed that bamboo tools could have been used extensively,[18] perhaps in a way that substituted for hand axes, but evidence of them would not likely be found in fossil sites. The problem with this explanation in the case of Longgushan is that bamboo today grows only in the south of China, nowhere near Dragon Bone Hill. Inferences from the fossil fauna also make it doubtful that southern Chinese flora ever extended up to northern China. The panda (genus *Ailuropoda*), which lives on bamboo and is common to southern Chinese Pleistocene sites, is unknown in northern China. Although quite a few fossil trees have been identified in Longgushan, bamboo and similar tropical species are not among them.

In the 1940s archaeologist Hallam Movius first proposed a line of de-marcation that separated the hand ax–containing sites of Europe and Af-rica from the Asian sites that lacked such artifacts.[19] The so-called Movius Line was located in Central Asia and extended down into the Arabian Peninsula. The line's location has always defied a reasonable explanation, but the general opinion has been that it represented a cultural division line or an ecological boundary. Hominids to the west were thought to have used hand axes, and hominids to the east used chopping tools. Recent archaeological research, however, has confirmed that though less abundant than in African sites, hand axes are found in China.[20] Sometimes, as in the case of Longgushan, there can be a mineralogical reason that hand axes have not been made. In other cases the reason is less clear. But underlying the entire issue is the fact that we still do not even understand what a hand axe was used for. Perhaps Asian *Homo erectus* did not appreciate its advantages either and found that a single-edged chopping tool did the job just fine.

Archaeologist David Hopwood has recently analyzed *Homo erectus* stone tools and the raw materials from which they were made.[21] He measured complexity of tool manufacture and the spatial aspects of sites, and from how far stone was brought to make tools. His findings show that early *Homo erectus* was quite similar in both Africa and Eurasia. About eight hundred thousand years ago in Africa *Homo erectus* sites became highly clustered, a pattern that Hopwood believes indicates a greater degree of social organization and social interaction. In Asia, however, he found a different pattern of regularly spaced sites that implied to him no substan-tial social interchange and perhaps even avoidance in the period between eight hundred thousand and six hundred thousand years ago. Asian tools show much more local derivations than African ones, while at the same time African tools were significantly more complex and raw materials there were being transported long distances. Implications of these intriguing findings remain to be investigated.

The stone tools that *Homo erectus* wielded at Longgushan were not its state-of-the-art technology. Fire was. If we imagine what it might have been like, eking out a life in Ice-Age China with only stone tools in or near a cave with large predators, survival seems an iffy prospect. Add fire to the equation, however, and we might feel that chances for survival were significantly enhanced. Indeed, paleoanthropologists have for many years generally believed that fire was a prerequisite for hominids to colonize the higher, colder latitudes of Eurasia as they expanded out of Africa.

A corollary of this idea is that the first evidence of fire should be found in early archaeological sites in Eurasia. Research by archaeologists J. D. Clark and J. W. K. Harris in very early archaeological sites in Africa beginning in the 1970s challenged this idea.[22] Clark and Harris found areas of baked clay on the fossil savanna landscape of Koobi Fora and Chesowanja, Kenya, dated to an astonishingly early 1.7 million years ago, at the very dawn of the species *Homo erectus*. Recent research on the geochemical aspects of Clark and Harris's data by Ralph Rowlett of the University of Missouri has provided support for this controversial hypothesis.[23] We think the data are solid and we agree that fire was tamed early by hominids. But if that is so then the original use of fire by hominids must have been for other purposes than to warm themselves against the Ice-Age cold and to light the interiors of dark caves. Hominids likely used fire in their interspecies competition for food and space on the African savanna. Those lessons were simply extended to different species and different contexts in Eurasia.

The evidence from Longgushan is compelling in indicating that the cave was primarily a hyena den that *Homo erectus* occasionally shared with a number of other species of carnivores, birds, bats, and rodents. Fire was first and foremost a means by which *Homo erectus* could effectively compete with these other species and hold its own in the hurly-burly of Pleistocene China. Cooking was only a by-product of this adaptation and rendered certain foods that were difficult to process, such as a horse head, much more accessible and palatable. But it is likely that hominids, like their primate cousins the macaque monkeys (also at Longgushan), had a diverse enough diet without fire to survive. The warmth that fire provided was also likely a benefit in severe conditions, but the primary adaptation against Pleistocene cold was undoubtedly shelter, not fire. Small shelters probably date back to pre-fire-using *Homo*, again in Africa, and they would have been effective without an interior fire. In fact, without a chimney small huts can be rendered much *less* habitable by the smoke and soot from a fire.

The early advantage that fire gave to *Homo erectus* was a leg up on the competition with other species—a competition that was exacerbated by climatic changes accompanying the onset of the Ice Ages. In the next chapter we look at how both global changes in climate and local conditions at Longgushan affected the life of *Homo erectus*.

Hunter, Gatherer-Hunter, or Scavenger?

Homo erectus used to be thought of as the first hominid to have engaged in big-game hunting—running down and dispatching animals larger than themselves. Part of this opinion was rooted in the Western cultural memory that hunting was primitive, predating agriculture, and was the way our early ancestors made a living. The evidence of this mode of life was the large animal bones found at such archaeological sites as Torralba and Ambrona in central Spain, investigated by paleoanthropologists F. Clark Howell and Leslie Freeman beginning in the 1960s.[24] Here mammoth bones were found in association with *Homo-erectus*-aged stone tools. Hominid bones were never found, but the association of big-game hunting stuck with *Homo erectus* nevertheless.

A new generation has questioned the big-game hunting interpretations of Torralba-Ambrona and other early hominid sites. These archaeologists point out that the elephant bones and stone tools merely showed that hominids had cut up large animals and presumably eaten them. It said nothing about how the hominids had come into possession of the carcasses in the first place. Among the most powerful tools that these researchers used was the scanning electron microscope (SEM), with which they examined the surfaces of the bones at archaeological sites. SEM photomicrographs became important for distinguishing the various marks left on bones—from shallow parallel scratches resulting from trampling by antelopes to deep U-shaped grooves made by the teeth of carnivores, to the repetitive closely spaced gnaw marks of rodents, to the V-shaped and sharply incised cut marks left by hominid stone tools. Analysis after analysis of early hominid sites showed that cut marks almost invariably overlay bite marks, indicating that carnivores had eaten part of the meat first, and had presumably hunted down the animal initially. Zhoukoudian fit into this pattern of reinterpretation of big-game hunting. Both Lewis Binford's bone damage studies and our own have supported the interpretation that *Homo erectus* at Longgushan were scavengers and not hunters.

Support for the scavenging hypothesis comes from an unlikely source. Parasitologist Eric Hoberg and his colleagues at the U.S. Department of Agriculture studied the three species of tapeworms that infect the human species. In comparing them to all the known species of mammalian tapeworms, they found that they were closest to the tapeworms that infest hyenas, felids (lions and tigers), and canids (dogs and wolves). *Taenia solium*, known as the "pork tapeworm," is the primary species infesting humans. It shared a recent common ancestor with the hyena-infesting species, and according to molecular studies, they diverged evolutionarily between 1.7 million (the beginning date of *Homo erectus*) and 780,000 years ago.[25]

Taenia saginata (the "beef tapeworm") and *Taenia asiatica* (the "Asiatic tapeworm") are other tapeworms that infest humans and, quite surprisingly, they also diverged at about the same time from the tapeworm that infests cats. Hoberg hypothesized that the tapeworms rode out of Africa in their migrating hosts (prey animals like pigs and antelopes) and were picked up first by early hominids "when they started eating more of the same worm-bearing meat as big cats and hyenas did."[26] Both the timing and the species involved make sense from the standpoint of the paleontological and archaeological evidence for scavenging from Longgushan. But the tapeworm data add substantially to the story.

Homo erectus can only have contracted infestations of these tapeworms, living in different intermediate hosts, if they first began to eat these new mammal species, or began to eat much more of them, at 1.7 million years ago. The fact that three separate and independent species of tapeworms from three separate mammalian species (presumably a pig, a bovid antelope, and an unknown Asiatic mammal) adapted to the hominid digestive tract at the same time strongly implies that meat from different species became a much more significant part of the hominid diet at this time. These data alone could be explained by positing hominids eating host prey animals and the infesting tapeworms diverging and adapting to live in the hominid digestive tract. But the problem becomes a multispecies ecological puzzle when we consider that carnivores were also eating the same prey species, thereby sharing their parasites. And not once, but thrice. How did the tapeworm species that began to parasitize hominids also share ancestry with the species that infested carnivores?

A solution to the puzzle may be to imagine the world from the standpoint of the tapeworm. Its eggs are eaten by a prey animal, such as a deer, and then hatch inside the digestive tract. Larvae burrow through the intestinal wall and enter the prey animal's bloodstream. Larvae then go into a quiescent, encysted state called a cysticercus and become embedded in the deer's muscle. There they wait to be eaten by a meat-eating species, such as a hyena (the definitive host), in whose digestive tract they can finally realize their potential and develop into adult tapeworms. It must be a lonely and risky life for most tapeworms, many of whom may never be rescued by a marauding carnivorous species from their larval state of suspended animation. And even if their intermediate host mammal is killed and eaten, imagine the tapeworm's disappointment when, finally freed from its fleshy tomb, it fails to survive in the intestines of a foreign definitive host such as *Homo erectus*. Natural selection would be expected to favor a tapeworm species that was adaptable and could survive in various common intestinal environments. It is most reasonable to conclude that three different tapeworm species, originally adapted to specific intermediate and definitive hosts, speciated about 1.7 million years ago to become generalized parasites who could take advantage of a new digestive environment—that of *Homo erectus*. This is independent evidence that *Homo erectus* had adapted to a diet that contained more meat than previously, that the species eaten were varied, and that a close ecological relationship with mammalian carnivores existed.

The tapeworm evidence adds significant detail to our understanding of *Homo erectus* and fire. A close ecological relationship between *Homo erectus* and large mammalian carnivores, which we know from other evidence also ate hominids, could only have been made possible by the hominid possession of fire. Fire would have been crucial in enabling hominids to obtain meat predictably by scavenging because it was only in this manner that hominids could ever have displaced larger, fleeter, stronger, clawed, and fanged competitors. And tapeworms yield yet another clue about the use of fire. Only when meat is eaten raw or very undercooked do tapeworm cysticerci survive to infest a human digestive tract. We may surmise that in many cases *Homo erectus* ate meat the same way that the carnivores from whom they scavenged ate it—raw. If meat was cooked at Longgushan, *Homo erectus* must have preferred it rare. Otherwise the tapeworm species would never have survived to adapt so well to the hominid digestive tract.

The picture of *Homo erectus* culture that Longgushan preserves is a primitive one, a world apart from our own. Nevertheless, *erectus* culture was a powerful adaptation for its time. Using rudimentary stone tools and a tenuous control of fire, and having a dependent scavenging relationship with dangerous large carnivores, *Homo erectus* did more than eke out an existence in Pleistocene. It flourished, multiplied, and expanded its range. The ecological conditions that attended this unlikely evolutionary transition help explain how and why these adaptations evolved. We next look at *erectus*'s world.

CHAPTER 6

The Times and Climes of
Homo erectus

The bullnecked and bullet-headed species *Homo erectus* was physically primitive enough to be a compelling human ancestor. But once the anatomical descriptions of the hominid, authored by Davidson Black and then Franz Weidenreich, were largely completed, paleoanthropologists (and the public in general, to whom "Peking Man" had become a household name) wanted more details. Like the forthcoming sequel to a novel or the next installment of a serialized movie, the latest research findings from China were awaited with eager anticipation. Many questions surrounding *Homo erectus* related to their exotic context. Just how old were these fossils? Did *erectus* live through the cold of the Ice Age? Where had they come from? By drawing on fields outside the traditional realms of physical anthropology and archaeology, we have been able to piece together much of when and under what conditions *Homo erectus* lived.

Questions about *Homo erectus* behavior have to be placed in an environmental context first. Did these hominids from the Ice Age live amid snow and ice, adapting like modern-day circumpolar peoples such as the Inuit, or were conditions less extreme? Did they stay in one place—for example, near Longgushan—most of the year, or did they move with the seasons? What sort of shelter and clothing would have been necessary? Most of these pressing questions were not answered in the lifetimes of the first investigators (and some are still not answered) because resolution has had to wait for sophisticated dating techniques and the integration of results from many different subfields of science. Only recently has it been possible to construct a firm contextual story for the Longgushan fossil hominids, and the most basic part of the story is the geological age of the site.

The Age(s) of Longgushan

By the early 1930s it had become apparent to the excavators that Locality 1 was a single big cave infilling. Sediments and bones had become incorporated in a more or less continuous process lasting tens or hundreds of thousands of years. Yet there were smaller discrete fossil deposits spread over Dragon Bone Hill that were unconnected with Locality 1. These were surveyed and designated as different "localities" (as distinguished from the "loci" within Locality 1 discussed earlier).

The relative ages of these various localities were bracketed by the species of fossil mammals discovered in them. This way of determining the age of fossil deposits is known as biostratigraphy, and it has been used since the dawn of paleontology. Locality 12, for example, located half a kilometer east of Locality 1, had in its inventory of species a more primitive monkey (*Procynocephalus*) than the species at Locality 1 (*Macaca robusta*). It was assigned an early Pleistocene age. On the other hand, the Upper Cave site (also known as Locality 26) yielded a fauna essentially identical to that of northern China today. It was assigned a latest Pleistocene age. In all, there are 45 localities at Dragon Bone Hill that span the time from the recent past back through the entire Pleistocene Epoch, and even into the epoch before, the Pliocene. These localities tell the story of the climatic and biotic evolution of China before and during the Ice Ages.

Locality 15 is the oldest known locality at Longgushan. It is unique among the localities in many ways. First of all, its sediments are river-laid sands, silts, and gravels. And the fossils from Locality 15 are virtually entirely those of freshwater fish. Locality 15 records a time when the land surface at Longgushan was lower relative to the Zhoukou or Ba'er River, because water inundated the cave when the river was high and during floods. Fish were swept into the cave and died where they were trapped. As the water evaporated, their skeletons were entombed in the enclosing river mud that eventually turned to stone and fossilized them. Locality 15 is dated to the Pliocene Epoch because the species of fish found there are the same as those found at other Asian Pliocene sites. The site has never been dated by absolute dating methods (techniques that can give an age in years) so its date is termed "relative." We estimate that Locality 15 is between three and five million years old.

Sometime after Locality 15 was deposited, Locality 12 formed. Many years are missing between these two localities because Locality 12 is early Pleistocene, dated by relative dating to less than two million years ago. During this span of time the Zhoukou River cut down into its valley, meandering in S-shaped turns as it did so. The bedrock of Longgushan was slowly uplifted, pushed up by crushing forces in the earth's crust and movement

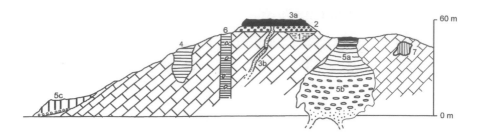

The other localities of Dragon Bone Hill. Above: Geological sketch (not to scale) of Dragon Bone Hill localities and sediments by Teilhard de Chardin. Below: View of Dragon Bone Hill from the north. Dragon Bone Hill is, like the rest of the Western Hills, made up predominantly of limestone originally deposited in ancient seas during Ordovician times (ca. 400 million years ago). The much more recent sediments were deposited in these uplifted, tilted, and faulted limestone rocks. The top of Dragon Bone Hill is composed of the oldest of these recent sediments—remnants of a Miocene and Pliocene cave system now almost entirely removed by erosion (1 = "Yellow Sands" with fish fossils; 2 = gravel; 3a = cave flowstone (stalagmite). Locality 12 (3b) was a fissure infilling of this early cave system in which many monkey fossils were found in the "*Cynocephalus* Gravels." Locality 13 (4) is an early Pleistocene deposit of red clay with many fish fossils, probably representing a lake deposit. Locality 1 (5a, 5b) represents early-middle Pleistocene deposits, which are the only sediments known to preserve *Homo erectus* fossils. Terrace deposits near the Zhoukou River (5c) may correspond in age to the Locality 1 sediments. Upper Cave (7) with late Pleistocene exposures can be seen adjacent to Locality 1. Locality 3 (6) is a middle Pleistocene deposit with few fossils.

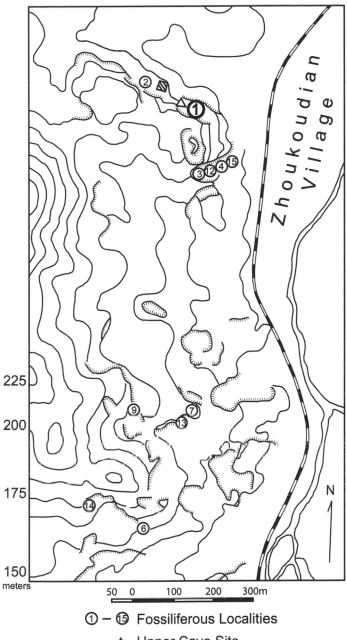

① – ⑮ Fossiliferous Localities

△ Upper Cave Site

Fifteen localities were originally established on Dragon Bone Hill. Eleven of these localities can be identified on this map. Only Locality 1 (in bold) preserves hominid remains, but the others provide valuable paleoenvironmental data for the conditions before and after *Homo erectus* frequented the cave at Locality 1. A total of 45 localities have now been established at Dragon Bone Hill. For views of other localities see previous figure.

along cracks or faults in that crust. The net effect of these geological events was to cause a much drier interior in the developing Longgushan cave. The cave began to be populated by terrestrial animals.

Pierre Teilhard de Chardin first noted the steady uplift of the Zhoukoudian region from the Pliocene to the present.[1] This uplift accounted for the major change in the fossil fauna from fish to terrestrial. Teilhard synthesized much of the information about the early Longgushan localities and produced a summary chart.

From the standpoint of our interest in human evolution, the uplift of the Longgushan cave and its trapping of terrestrial animal bones was a good thing. But the fact that the cave was now above the river's floodplain meant that river sediments would no longer come into the cave and cover the bones, fossilizing them. Sediments could now only come to enclose bones when deposited by groundwater seeping through the roof and walls of the cave—as in stalactites and stalagmites (flowstone), by rainwater runoff pouring in through the external openings of the cave, or by sediments blowing in from the outside. The deposits in Longgushan Cave preserve some flowstone, but the predominant type of sediment that fills the cave is breccia (Italian for "broken"), composed of fallen boulders and slabs of bedrock from the roof, cemented by washed-in silt, sand, and loess from the soil surface above.

Teilhard de Chardin, in a remarkable feat of reasoning—considering the data available to him—deduced that the times of deposition at Longgushan would have had to correlate to periods of increased rainfall (and decreased ice), which would have been during periods of relative warmth. The gaps in time between the various localities at Longgushan then were explicable in terms of cycles of sedimentation—sediments and fossils were deposited only during wet and warm periods of the Pleistocene. During periods when water was locked up in glaciers, and rainfall was scarce, there was no sediment washed into the cave. Teilhard correlated the periods of increased sedimentation (and fossil deposition) at Longgushan with the cycle of Pleistocene "interglacials"—the periods in between the glacial periods long known in European and North American geology.

Despite the prescience of Teilhard's inspired geological deductions, the history of sedimentation and therefore the ages of the fossils in the cave remained unproven until methods of absolute dating were applied to the cave. But this sounds much easier than it proved to be in practice. Too old for carbon-14 dating and lacking potassium-rich volcanic rocks that could be dated by the potassium-argon method, the sediments at Longgushan had to await the development and refinement of various absolute-dating methods that have only recently begun to yield consistent results.

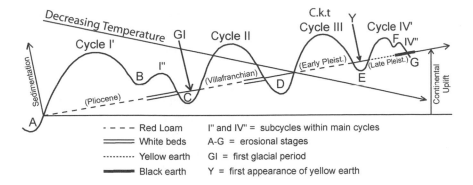

Teilhard de Chardin's concept of past climate in northern China. Sedimentation increased during warm periods ("interglacials"), with increased rainfall and melting glaciers, numbered I through IV. Glacial periods were times of decreased sedimentation, lettered A through G. Teilhard placed Zhoukoudian Locality 1 ("C.k.t.") in Interglacial Period III (his Cycle III), a warm and wet period. Superimposed on this cyclic climatic change was geological uplift of the sediments as well as a general trend toward colder conditions over time. Modern research has modified and refined this framework but much of Teilhard's model is still valid.

Nailing Down the Dates

Everyone knew that Longgushan and its fossils were old. But how old? Weidenreich thought that the Chinese *Homo erectus* were more evolved and more humanlike than Javan *Homo erectus*, and he speculated that they were more recent. Absolute dates could help resolve the time relationship between the hominids from China and those from Java. In addition, as more and more hominid fossils turned up in China in later years, the date of the cave near Zhoukoudian also became an issue. There were early *Homo sapiens* (more likely *Homo heidelbergensis*) fossils from China that rivaled the presumed ages of fossils at Longgushan. Could the two species have been contemporaries? And as earlier and earlier *Homo erectus* fossils were discovered in Africa, back more than a million and a half years, the dates for Chinese *Homo erectus* at Longgushan began to look more and more anomalous. Chinese scientists began an earnest quest for methods to date the cave sediments at Longgushan.

The basic premise of most absolute-dating methods is assessing the state of decay of a chemical element. That element can be a part of a fossil itself or it can be a component of the geological context of a fossil. Geochronologists—those scientists who research the age of the earth—have been very resourceful in finding elements to date. Many of the dating techniques use different forms of an element. These different forms of an element are called isotopes and they differ from one another in the number of neutrons in the nucleus (isotopes of the same element by definition share the same number of protons or "atomic mass" in the nucleus).

Carbon dating is the granddaddy of absolute-dating methods. Invented by Willard Libby in 1955 in Chicago, it measures an isotope of carbon, carbon 14 ("14" referring to the number of protons and neutrons in the atomic nucleus). The carbon 14 isotope is formed in the atmosphere from nitrogen 14 by ultraviolet radiation from the sun. A neutron collides with nitrogen, turning it into radioactive carbon 14 by releasing a proton as hydrogen and adding a neutron to the nucleus. Carbon 14, with six protons and eight neutrons, is found in the same amounts in all living organisms—just so long as they remain alive. When they die, no more carbon 14 is breathed in or ingested, and the amount of the isotope begins to disappear. It decays at a standard rate, losing its extra neutrons as the atom returns to more stable, lower-energy states. What is critically important for absolute dating is that elements decay at very predictable rates. Carbon 14 decays at a rate that removes one-half of the original amount in 5,700 years (a period of time known as its half-life).

Carbon 14 decays at such a rapid rate, in geological terms, that too little of the isotope is left after about fifty thousand years to yield a very accurate date with conventional carbon 14 dating. A newer method, potassium-argon dating, pioneered by Garniss Curtis and Jack Evernden at Berkeley in the 1960s, extended the time scale of absolute dates to millions of years. The era of ever-earlier African hominid discoveries began, calibrated by potassium-argon dates on the volcanic (and potassium-rich) rocks that bracketed the fossils. But back in China, and at other hominid fossil–bearing cave sites the world over, no revolutionary new dating methods existed. As the early hominids of Africa assumed the starring roles, the proverbial "cavemen" were pushed to the side in the new and seemingly ever-older narrative of human evolution.

Uranium-series dating has now allowed absolute dating of cave sediments in the time span of approximately three hundred thousand to more than a million years ago. The theory of uranium dating has been known since the late nineteenth century. But accurate determination of uranium's isotopes and the possibility of loss of decayed products to the enclosing rock have made application of the theory difficult. New advances have now made uranium dating reliable. A number of isotopes of uranium occur in nature, and each decays at a known rate. Uranium 238 decays to uranium 234 slowly (its half-life is 451 million years) whereas uranium 234 decays to thorium 230 much more rapidly (with a half-life of 245,000 years). Uranium dissolves easily in water and is also deposited in minerals precipitated from ground water, such as cave flowstone, also known as travertine, which is composed of calcium carbonate. Because bone and teeth are of similar mineral composition, it is also possible to determine the ages of fossil bones and teeth by uranium dating.

In 1985 Chinese scientists led by Shusen Zhao were able to measure the amounts of uranium 238, uranium 234, and thorium 230 in a sample of fossils from Longgushan using a mass spectrometer.[2] This is an apparatus that "catches" and counts electrons of specific energies and thereby accurately determines the quantities of isotopes in a sample. Knowing the numbers of isotopes and their decay rates, one can calculate the age of the sample. Scientists determined two ages for Layers 1 through 3 in the Longgushan cave—230,000 years old and 256,000 years old—but there was more variation in the individual determinations than was expected. This implied that some specimens had lost some of their decay isotopes, probably due to weathering of the bone or some other "diagenetic" change, and thus appeared too young. In a study six years later, S. Yuan and his colleagues reported on a more intensive study of carefully selected and prepared bone from Layer 2 that gave an older and apparently more accurate date (with less scatter of the determinations) of 290,000 years.[3] A mean age with a tighter cluster of measurements around the mean, they hoped, meant a more accurate age, but it could also simply mean that all samples sitting buried in the same sediments for hundreds of thousands of years had lost some of their isotopic-decay products to the same extent, not an unlikely proposition. Then in 1996 G. Shen and his colleagues used a new and more accurate measuring technique, thermal ionization mass spectrometry (TIMS)—a method of "step-heating" the samples to release their isotopes.[4] They could thereby determine the oldest uranium isotopic-decay products from Layer 2 travertine. They determined the true age of the rock to be much older than the previous determinations on fossilized bones—410,000 years. The effect of the new dates was to push back the age of Peking Man some two hundred thousand years.

The new dates were substantially older, and from the top of the Longgushan sediments. Other absolute-dating techniques were tried in order to confirm the old dates and to attempt to ascertain the age of the lowest sediments of the deposit. Paleomagnetism, a dating method based on the surprising phenomenon of Earth's flip-flopping of magnetic north and south during its geological history, proved important in defining the lower age limit of the cave deposit at Longgushan. Sediments record the orientation of magnetic north microscopically in the orientation of their sedimentary particles. Study of the sediments at Longgushan by F. Qian and his colleagues in 1985 revealed that the boundary between the Brunhes Epoch ("normal" polarity, during which a magnetic needle points north) and the older Matuyama Epoch ("reversed" polarity, during which a magnetic needle points south) occurs in Layer 14 at Longgushan.[5] This boundary is dated as a worldwide event at 780,000 years ago, so the first fossiliferous strata at Longgushan, in Layer 13, are almost this old. The oldest hominid fossils occur in Layer 10, where the first complete skull (Skull III) was found at Locus E, and are probably about 110,000 years younger. This estimate is based on the rate of sediment accumulation in the cave. It establishes the oldest *Homo erectus* at Longgushan to be about 670,000 years old.

The accurate dating of Longgushan has made it possible to piece together the climatic history and ecology of *Homo erectus* in China. *Homo erectus* occupied Longgushan Cave intermittently from about 670,000 to 410,000 years ago.[6] This was a span of time that we can now confirm as the middle Pleistocene—a period within the Ice Age when climates fluctuated between being very cold and being as warm as today.

Weather Report from Longgushan

Teilhard's theory of sedimentation at Longgushan has stood the test of time, and it is still accepted in large part by the modern international team of geoscientists who have worked on the site.[7] Teilhard noted changes in sediments and fauna related to geological uplift of the area and to a trend toward increasingly colder climate. But there were other changes in the sediments and in the fossils that could not be explained by altitudinal changes and the impending Ice Age alone. Smaller-scale fluctuations from warm to cold occurred within the general trend toward colder conditions. The implications for understanding *Homo erectus* behavior in this record are profound.

In a masterful synthetic study of all paleoclimatic data from the Longgushan cave, in 2000, Chinese geologists Chunlin Zhou and his colleagues correlated the climatic fluctuations that Teilhard had first seen with the global curve of climate change in the Pleistocene Epoch.[8] For the first time it was possible to correlate specific layers in the cave (and their enclosed hominid fossils and stone tools) with a detailed reconstruction of the environmental conditions in northern China at the time. For their "synthetic climatic index" Zhou and his colleagues incorporated such measures as weathering of sediments at each stratigraphic layer. For this index they counted individual grains of quartz sand in a standardized sample and compared this number to the number of grains of the mineral feldspar. Higher percentages of quartz show more washing in of sand from the ground surface than bedrock-derived feldspar, and these peaks of the index correspond to warmer and wetter climatic conditions. During periods in which there was less quartz than feldspar, high proportions of cold-adapted plants, such as the grasses *Artemesia* and *Selaginella*, were seen.

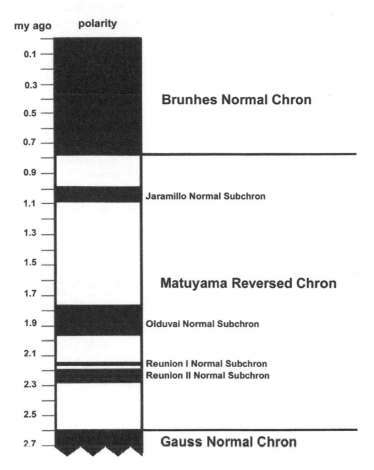

The global paleomagnetic stratigraphy for the past 2.5 million years. The paleomagnetic signature of sediments at Longgushan have assisted in dating the deposits. "Normal" refers to those times in earth history when magnetic polarity was north-facing, as today. "Reversed" refers to periods when magnetic polarity was south-facing. The boundary between the Brunhes Normal and the Matuyama Reversed Chrons occurs in Layer 14 of Locality 1, indicating that these sediments underlying the hominid-bearing deposits of the site are over 780,000 years old. my = millions of years.

Isotopes also play a part in reconstructions of past climate. In 1985 Xie and his colleagues analyzed the relative amounts of the elements barium and strontium in sediments from Locality 1.[9] When there is a high ratio of barium to strontium, the climate can be inferred to have been wetter and warmer because, as sediments weather, strontium is leached out and lost from the soil. These data were also used by Zhou and his colleagues to put together their climatic index. The peaks of graphed strontium/barium ratios through time fit nicely with the other measures of climate from

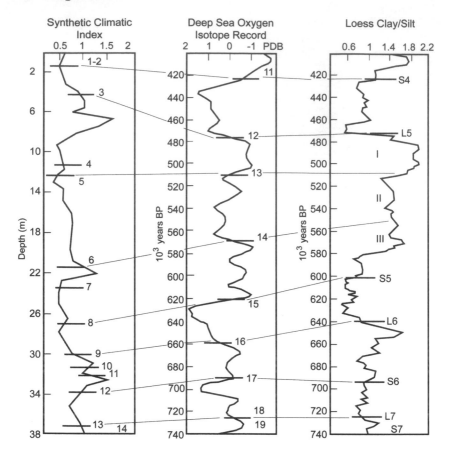

The Pleistocene Epoch, known colloquially as the "Ice Age," was actually a time of fluctuating conditions—from very cold to very warm. These diagrams, constructed by paleo-
climate researchers Chunlin Zhou and his colleagues (2000), record the climate in northern
China during the time that *Homo erectus* lived near Dragon Bone Hill. A Synthetic Climatic Index integrates data from weathering of sediments. The Deep-Sea Core records
temperature changes as a function of oxygen isotopes. "Loess" accumulates during periods of cold, glacial conditions. As a composite, these data show that *Homo erectus* adapted
to changing conditions, living near Dragon Bone Hill in warm interglacial conditions
and probably migrating to the south during glacial conditions.

Longgushan, such as the record of loess. We will return to the climatic
story that isotopes tell when we correlate Chinese *Homo erectus* with global
patterns of change and migration in chapter 8.

During the relatively "good times,"—the interglacial periods, when rainfall was plentiful, temperatures habitable, and game and plant foods abundant—*Homo erectus* lived around, and even in, the Dragon Bone Hill cave.
We know this by the records of stone tools and cut marks left on bones in

the cave. During cold, "glacial" periods, the wind swept off the Mongolian steppes carrying the dry, ice-ground loess, and the large mammals went south. Only arctic-like small mammals and cold-adapted plants are recorded in the Longgushan sediments during these periods. Hominid fossils are rare or nonexistent during the cold phases. Where did they go?

The implications of the paleoclimatic record at Longgushan are that *Homo erectus* was still largely a tropical species, migrating south to be with the warm-adapted southern China fauna and flora (species like the panda and the bamboo tree), when it became cold, and returning to northern China during the warm periods. The Qinling Mountains cordon off northern China from southern China in an east–west wall that created a major physical and climatic barrier even then. These mountains would have essentially shielded southern China from the glacial winds blowing from the north during cold periods of the Pleistocene. Southern China would have served *Homo erectus* populations as a refuge when northern China was too cold to be habitable.

The large-scale migratory pattern that we deduce for *Homo erectus* emphasizes the adaptive limitations of this species compared to modern humans. *Homo sapiens* groups such as the Inuit, Saami, Paleo-Indians, and ancient Siberians all adapted to frigid climates in the far north, building effective shelters, successively finding food by hunting, herding, and gathering, and keeping warm with well-made clothing and fire. It has also recently been suggested that dogs were domesticated to help *Homo sapiens* hunt and fend off predatory hyenas.[10] We must conclude that the essential components of these types of adaptations were missing in *Homo erectus*.

Exactly Where Did *Homo erectus* Live?

If *Homo erectus*'s life was unlike modern humans' lives in the far north, then what was it like? Did these proto-people live in caves or not? And if they did, how did they defend themselves from the large carnivores that certainly did live there? If they were unable to harness fire effectively, what was their relationship to it?

We saw previously that *Homo erectus* at Longgushan did not apparently gather, eat, and sleep around a central point inside the cave and keep the fires stoked with wood from outside. Otherwise, we would see the telltale phytoliths and silica-rich residues characteristic of hearths in the cave sediments. We now also know that *Homo erectus* apparently migrated to the warmer south during the cold phases of the Pleistocene. From these two pieces of evidence we can deduce that *Homo erectus* followed the pattern of land use characteristic of its ancestors in Africa—camping in the open, probably with lightweight shelters built of tree branches and anchored by stones set in the ground. So far such a campsite has not been discovered in or around Longgushan, but we can be relatively certain that this is where the hominids whose remains ended up inside the cave spent most of their time.

The sediments of Longgushan cave give us no clue that hominids camped and lived there long-term. Stone tools and their marks on fossil bone are there, and evidence of episodic use of fire is there—both attesting to the presence of *Homo erectus*. But the archaeological pattern is more suggestive of a commando raid than of the comfortable cave home so often invoked by theorists of old. We believe that hominids armed with stone tools and weapons, the primary one of which—fire—they still did not fully control, entered Longgushan cave to pilfer meat from the resident carnivores. Perhaps bringing in dry brush from the outside, they torched the cave, setting fire to the dry guano and scaring off the hyenas, lions, wolves, and bears long enough to preempt their kills.

An interesting study on the paleoecology of Pleistocene sites in Africa provides an important although indirect confirmation of this model of *Homo erectus*'s ecological behavior. Lillian M. Spencer of the University of Colorado at Denver carried out a study of the savanna-adapted antelopes living in the period during which *Homo erectus* first evolved.[11] She found that grazing species adapted to secondary grasslands became prevalent about two million years ago. Secondary grasslands are maintained by fire that is caused by the increasing aridity of climate and, at least today, also by human fire setting. We hypothesize that fire became an important ecological tool for *Homo erectus*, a means by which the species could extend its optimal environment and its control over other species. The adaptation became powerful enough to allow the species eventually to spread out of Africa and into Eurasia.

Physical changes accompanied *erectus*'s migration across the Old World. Body size increased and legs increases in relative length. Longer lower limbs meant that the stride became significantly longer. More ground could be covered in a single day of foraging for food over open, fire-maintained grasslands.[12] Although the fossil remains from Longgushan are not complete enough for us to make this deduction directly, the much more complete skeleton of the "Turkana Boy" from Kenya has demonstrated this fact of *Homo erectus*'s anatomy.[13] Thus, ecology relates to anatomical change. In an interesting paper that compared the long-legged patas monkeys to their short-legged vervet monkey cousins, Lynne Isbell and her colleagues suggested that efficiently covering a large area to find food was the critical element in effecting this anatomical change.[14] Comparing the patas monkey to *Homo erectus*, these authors suggested that the same ecological and evolutionary forces worked to increase hominid leg length. *Homo erectus* literally walked for their food.

The Evidence of Peking Man's Diet: Brain and Hackberries, Anyone?

Ecology can tell us much about the behavior of *Homo erectus*, particularly its dietary behavior. Vegetable foods were undoubtedly important to this still semitropical species, but protein from meat was also demonstrated to be a major aspect of *erectus*'s cuisine. Fire was therefore probably important. As we know from the archaeological data, some of the meat that *Homo erectus* scrounged in the cave and cut off of old kills of carnivores was less than fresh, and natural selection may have favored a taste for seared steak at this time. Certainly, as we now know from genetic studies, carnivore tapeworms had already colonized the hominid digestive tract. Any behavior that reduced this type of parasitism would have been beneficial to the species.

What of the rest of the diet of *Homo erectus*? Binford and Stone's discovery of the remnants of roasted horse heads at Locality 1 is important because it shows that organ meat—in this case, the brain—was also eaten. The brain is a fatty (lipid-rich) organ that mammalian carnivores tend to relish. We have already seen that hyaenids at Longgushan expended substantial effort to get at the brains of the hominids who fell into their clutches.

Ralph Chaney was a paleobotanist from the University of California at Berkeley who worked at Zhoukoudian in the 1930s. He discovered and subsequently identified an abundance of seeds

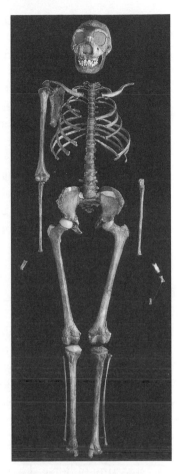

Homo erectus from Nariokotome, west of Lake Turkana, Kenya, nicknamed "Turkana Boy." This almost complete skeleton is dated to 1.55 million years ago and is the most complete early evidence of the species. We term Turkana Boy's species *Homo erectus ergaster*, to distinguish it from the later *Homo erectus erectus* from Longgushan and Java.

belonging to the hackberry tree, *Celtis* sp.[15] Fruits of the *Celtis* tree are eaten by primates in the wild, and Chaney surmised that the seeds found in the cave were remains of *Homo erectus* meals. But an equally plausible scenario is that a hackberry tree grew near the entrance of a vertical opening into the cave, and berries simply fell in as they became ripe. Another possibility is that birds that roosted in the cave ate the berries, and deposited the seeds in their guano on the cave floor. So, unfortunately, we cannot say for sure from the fossil evidence that *erectus* dined on hackberry seeds, even if they may have been a regular or seasonal component of their diet.

Another way to assess what *erectus* ate at Longgushan is to reconstruct the plant life during the times that the species lived in the area. Pollen in the sediments has been identified and can give a general idea of what plants grew there. Food plants that modern Chinese primates eat are today found only in southern China and were likely present around Longgushan only during the warm interglacial periods. The replacement of these woody and forest-adapted species by the grasses and sedges of the glacial periods may have been a major impetus for *Homo erectus* to migrate as well.

Ecological Relationships with Other Animals

The cave at Longgushan preserves evidence of the close association of hominids and a number of other animal species. Large mammalian carnivores, a number of mammalian herbivores, and many bird species are prominent members of this ecological community.

Cut marks on ungulate bones are unambiguous indicators of hominids' ecological relationship with such deer-like species as Gray's sika (*Pseudaxis grayi*) and the giant elk (*Megalocerus pachyosteus*). Tongue was clearly a commonly eaten body part. And we have noted the eating and apparent roasting of horse. There are also what appear from published photographs of fossil specimens that are now lost to be cut marks on bones of rhinoceroses, elephants, and pigs. Smaller animals that may have been eaten include tortoises and birds. Past authors have theorized that *Homo erectus* hunted these animals. But how did *erectus* really obtain these food sources?

Archaeology shows that *erectus*'s stone-tool technology was minimal. With no long-distance hunting implements like the spear, or the much more advanced atlatl and bow, *erectus* bands would have had a very difficult time dispatching an animal as large as a woolly rhino. Smaller and slower animals such as tortoises may have been preferred prey. Larger prey was likely scavenged.

The glacial-interglacial population movements that we have postulated for *Homo erectus* suggest that hominids may have followed migrating herds of herbivores as they moved seasonally. This idea was proposed some years ago by ecologist Norton Griffiths and archaeologist Mary Leakey for African early hominids, but it may well have been a pattern of ecological behavior shown by Chinese *Homo erectus* as well. We will return to this important model of hominid dispersal during the Pleistocene in our discussion of global population movements and the origin of *Homo erectus* in chapter 7.

Environment and ecology provide us with a good basis on which to investigate the behavioral traits of *Homo erectus* that made life in Ice-Age China possible—intellectual ability, manual dexterity, and speech. In the next chapter we turn to these most human of *erectus* traits.

CHAPTER 7

The Nature of Humanness at Longgushan: Brain, Language, Fire, and Cannibalism

When Davidson Black lay dying at his lab bench in the wee hours of March 24, 1934, the last sight that his eyes beheld was an evolutionary sequence of *Homo erectus* to *Homo sapiens*—the skulls from Longgushan that he had laid out before himself. Yet the cave had not yielded to Black exactly what he had imagined it might—hominids, yes, but these hominids were much more anatomically primitive than what anthropologists in the 1920s had conceived a human ancestor to be. Black's *Sinanthropus* did not have the expanded braincase and globular skull of *Homo sapiens* or of the spurious Piltdown Man, but Black still firmly believed it to be humanity's ancestor. Perhaps that was the symbolism of his dying message to us—here I die amid the bones of my ancestors.

Bones of anatomically modern *Homo sapiens* from the Upper Cave at Longgushan showed the presence of the presumed descendant of *Homo erectus*, discovered remarkably in the very same site, and archaeology provided further evidence of behavioral evolution—advanced humanlike behavior, tools, and fire—in *Homo erectus*. There are other behaviors that have been difficult to interpret and controversial as regards the humanness of *Homo erectus*. We now examine the data for whether the species could communicate by spoken language and whether it was cannibalistic.

The Anatomy of Speech

Human speech is a remarkably complicated cooperation of our brain, mouth parts, tongue, voice box, and breathing apparatus. Most anatomists and paleoanthropologists studying human speech and its origins have focused on the brain. After all, it is the brain whose significant enlargement in humans presages the evolved ability to communicate with spoken language. The brain is large in *Homo sapiens*, who we know can speak, and relatively small in apes, who we know cannot speak (although they are capable of some symbolic communication). *Homo erectus* falls between the two in brain size. Could the *erectus* people who lived around Longgushan in the Pleistocene speak or not?

The size of the *Homo erectus* brain is estimated by anthropologists to have been between 950 and 1,200 cubic centimeters. Some modern people, known as microcephalics, have brains this small. Can they talk? The answers vary. Many microcephalics, if they survive infancy, are severely mentally retarded and have no effective use of spoken language.[1] Others are normal. A rare few are even above normal. Anatole France, the nineteenth-century French author and playwright, had a very small brain (reportedly weighing 1,040 grams, which we can take as approximately equivalent to 1,040 cubic centimeters, and thus in the middle of the brain-size distribution for *Homo erectus*),[2] but that fact somehow did not impair his fluent use of language. He was awarded the Nobel Prize for Literature in 1921. Brain size then, by itself, does not provide us with an ironclad argument for or against *Homo erectus*'s capabilities for language.

Davidson Black first weighed in with his opinion that the *anatomy*, not the *size* of the *Homo erectus* brain, was evidence that the species could talk. By 1933, when Black published his opinion,[3] he and his colleagues had gathered data that they believed showed that *erectus* hunted, used fire, and made a diversity of stone tools—behavior much like that of later humans in the archaeological record, such as Neandertals. Black pointed to the impressions made of the inside of the fossil skulls from Longgushan to infer the shape and form of the *erectus* brain. In particular he noted that the areas on the side of the brain in what would have been the frontal and temporal lobes seemed to be enlarged in *Homo erectus*, like in later humans and unlike in apes. This region of our brains contains language centers (termed "Broca's Area" or the inferior frontal gyrus, and "Wernicke's Area" or the superior temporal gyrus). As they have expanded in human evolution, these gyri have formed a prominent fold between them—the so-called Sylvian Fissure (also known as the lateral sulcus, separating the parietal and temporal lobes of the cerebrum). Black opined in 1933 that *Homo erectus* showed a humanlike form of the Sylvian Fissure and thus could speak. Many anthropologists since have tended to agree,[4] despite their reservations about inferring such an important behavior from doubly indirect evidence—not only is the inside of the skull an imperfect reflection of the outside of the brain, but brains show quite a bit of variation from one to another. It is possible to have a chimp brain with a partial, humanlike

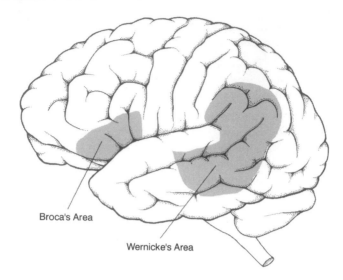

Broca's Area

Wernicke's Area

Speech areas of the brain. The outside layers of the cerebral hemispheres of the human brain are known as the cerebral cortex. From studies of modern-day patients with brain injuries, two areas—Broca's Area in the frontal lobe and Wernicke's Area in the temporal lobe—are known to be essential for language production and comprehension. The *Homo erectus* brain was smaller and lower than modern humans' brains, suggesting that Broca's and Wernicke's Areas were not developed sufficiently for the species to speak in a human manner.

Sylvian Fissure, as well as a human brain belonging to a well-spoken and fluent individual that shows only an incipient, apelike Sylvian Fissure. We cannot infer language ability with certainty from the brain endocasts of *Homo erectus* from Longgushan and other sites. Are there other anatomical clues that might tell us something about the species' ability to talk?

Language-Related Anatomy Outside the Brain

The idea that apes' brains are capable of sophisticated communication but that their peripheral anatomy does not allow them to speak has underlain the many studies in ape sign language. Washoe, the chimp, and Koko, the gorilla, for example, can both communicate in simple sentences by signing with their hands, but they cannot articulate words. By looking at the anatomical differences between humans and apes and comparing these findings with the fossil record, paleoanthropologists have come up with some interesting insights into the evolution of spoken language.

The anatomical position of the voice box, the larynx, has a lot to do with whether a hominoid can speak or not. Jeffrey Laitman at Mount

Sinai Medical School in New York and his colleagues have shown that the larynx is located high up near the base of the skull in apes and modern infants.[5] In adults the larynx descends, allowing a column of air above it to be manipulated to produce the controlled verbalizations of language. Laitman looked at fossil hominids to determine when in human evolution speech may have begun. He inferred the position of the larynx from the degree of flexion of the base of the skull—open and with an obtuse angle between the underside of the face and the base of the back of the skull in apes, and acutely angled in modern people. All the earliest hominids in Laitman's analysis are apelike in their skull base flexion and thus incapable of speech. Early *Homo sapiens*, now termed *Homo heidelbergensis*, were the first hominids to show enough basicranial flexion for Laitman to accept that they could speak. This analysis puts the Longgushan people on the nonspeaking side of the linguistic divide in human evolution.

The tongue also has a lot to do with language production. Richard Kay and his colleagues at Duke University looked at the size of the nerve to the tongue (the hypoglossal nerve, or cranial nerve 12) in apes and modern people and determined that it was relatively larger in people.[6] If the motor nerve to the tongue muscles is relatively larger, a greater number of nerve fibers must be coursing through it to innervate the tongue and control its fine movements. Since ape and human tongues are nearly the same size, and the other activities of the tongue—helping manipulate food while chewing, closing off the throat while swallowing, and getting out of the way while breathing—are the same in all hominoids, Kay and his colleagues reasoned that the relative increase in the size of the human hypoglossal nerve must be related to the fine motor control needed for language. The bony hypoglossal canal in the base of the skull must be preserved in fossils for this trait to be analyzed. Kay and his colleagues concluded that the australopithecines and earliest *Homo* from Africa had small, apelike hypoglossal nerves and thus were incapable of language. Earliest *Homo sapiens* had the enlarged hypoglossal nerves characteristic of modern people, and according to this analysis were the first to speak. All of the Longgushan skulls, as we have seen, lacked the skull base and thus the hypoglossal canal, and could not be included in Kay's study, but other *Homo erectus* fossils had small canals like their ancestors. Kay's conclusions were called into question by a study conducted by David DeGusta and his colleagues in 1998, who contended that they did not see the same increase in size of the hypoglossal canal from australopithecines to *Homo*.[7] However, the sample sizes are small and the postulated changes in canal diameters are difficult to measure precisely. Further study is needed, but for the moment there is some evidence that tongue anatomy tends to agree with laryngeal anatomy in indicating that *Homo erectus* was incapable of speech.

A final and unexpected contribution to the anatomy of speech came from the discovery and analysis of the early *Homo erectus* skeleton from the northern Kenyan site of Nariokotome. A study by anatomist Ann MacLarnon surprisingly revealed that this young male adolescent had a vertebral canal only three-quarters the size expected for a modern boy of the same age.[8] The vertebral canal is formed by the holes that run through vertebrae stacked one on top the other in the spine, and in life it contains the spinal cord. Overall body size, however, was within the range of modern humans. MacLarnon's controversial interpretation was that *Homo erectus* did not have the fine control of its respiratory chest muscles needed for language. Although this study agrees with other anatomical indicators of *Homo erectus*'s language ability, there are also a number of other nerve cells that run through the upper spinal cord, such as nerves for motor control to the hand and arm. The small diameter of *Homo erectus*'s upper spinal cord could as likely be related to less motor-nerve innervation to the hand and arm as to a decreased innervation to the muscles running between the ribs, used as accessory muscles in breathing. It is important to remember that the diaphragm, the major muscle used in breathing, is innervated from spinal levels between the third neck vertebra and the fifth neck vertebra, thus higher than the preserved fossil vertebral column from Nariokotome (which starts at the seventh cervical vertebra). Thus, we believe that citing the narrow vertebral canal of *Homo erectus* is a weak argument for the inability of the Longgushan people to speak, and that it may also (or instead) point to a lack of manual dexterity in *Homo erectus*, a topic to which we now turn.

Dexterity, Toolmaking, and Language

Paleoanthropologists have for many years made the equation between and among the ability to make stone tools, lateralization of the brain, and the ability to use language. *Homo erectus* clearly made and used stone tools, and thus for many years the species has been considered capable of speech, albeit at perhaps some decreased level of function. Anthropologist Grover Krantz, for example, made the intriguing but ultimately untestable suggestion that *Homo erectus* youths may not have learned to talk until adolescence.[9]

Our ideas about speech being the quintessential human attribute have also evolved. Anthropologists have for years been influenced by an idea of culture that was all or none. Many of the major cultural anthropologists of the last century, such as Alfred Kroeber and Clyde Kluckhohn, believed that cultural origins occurred at a specific critical point, like the boiling point of water, rather than a more gradual evolution through time. The "critical point" viewpoint was recently resuscitated by archaeologist Richard Klein who postulated a gene mutation fifty thousand years ago that suddenly made language possible—a linguistic "hopeful monster."[10] Tool use, documented by archaeologists, has been a more generally agreed-upon sign that culture has appeared. *Homo erectus* was for decades the earliest human ancestor found to have abundant stone tools (early australopithecines do not have stone tools), and thus in this view *erectus* was hailed as the first bearer of human culture. By inference then they were the first hominids capable of speech. Much has been learned to cast doubt on this view.

The language abilities of chimpanzees have been paramount in showing that symbolic communication need not involve speech. Chimps can, for example, use a red triangular plastic chip to mean "water" without ever uttering a word, which of course they cannot do. The gap between ape and human communication has narrowed significantly as more research has been carried out on chimpanzees, both in the laboratory and in the wild. Chimps may well represent a near-mute form of hominid—capable of symbolism but not of speech. Some primatologists even contend that chimps bear "culture," but if so it must be a very primitive form, without language.

The earliest tools have also shown that culture did not evolve all at once. Sileshi Semaw, now of Indiana University together with Jack Harris of Rutgers University and colleagues have discovered the very earliest stone tools at 2.6 million years ago in Ethiopia, and they are little more than smashed pieces of quartz with sharp edges. We only know that they are tools because of where they were found and their association with fossil bones having cut marks made by the tools. Deliberate flaking of stone need not be invoked to explain these tools. Hominids could have easily thrown lumps of quartz forcefully on the ground or against other rocks to make these stone tools. We can imagine chimps engaging in this sort of behavior, and indeed a recent discovery of chimpanzee stone artifacts suggests that the earliest hominid stone tools are little different.[11]

Homo erectus was a species in which natural selection increasingly emphasized brain growth. We deduce this from the fact that fossil skulls show increasing cranial capacity through time. We infer that an increasing brain size would reflect that more neurons and interconnections within the brain were developing to deal with environmental and interspecies challenges to survival. Language, manual dexterity, and increasingly lateralized functioning of the brain may all have been involved. But at the same time as the ballooning *Homo erectus* head was evolving to accommodate more brain, natural selection was also conferring on the species defensive head armor as postulated in chapter 3. These two adaptive trends, working in one sense antagonistically, must both have played a part in the composite adaptive evolutionary biology of *Homo erectus*.

Our significantly better fossil and archaeological records make it clear that toolmaking ability, fire use, and language must be decoupled in terms of their times of appearance and perhaps even their functional relationships to one another. That *Homo erectus* made stone tools is beyond dispute. Their using fire or making fire before the first hearths—dated in Europe at 230,000 years ago—is still contested by some archaeologists. We, on the other hand, are persuaded by the early evidence for fire and do not think hearths need be present before fire can be accepted as an important part of *Homo erectus*'s adaptation. Stone tools and fire are major cultural advances and we believe that they tie in importantly to increases in brain size and intellectual capacity. On the other hand it is unlikely that *Homo erectus* produced and used speech in a modern human manner because the anatomical parameters that we have adduced from its skull indicate an overall pattern unlike speech-producing *Homo sapiens*. Like other higher primates, *Homo erectus* certainly used a rich repertoire of verbal communication and, if we knew more about it, we might term it "protolanguage." There may have been a mixture of vocalizations and gestural language, à la American Sign Language, and even singing, but at present we do not know. And, unfortunately, we have not yet conceived a convincing way of finding out.

Summarizing what we believe that we know about *Homo erectus*'s behavioral abilities, we can say that the species made stone tools, used fire as an ecological tool if not as a primary domestic focus (as indicated by the lack of hearths), and lacked human language. We next turn to one of the most contentious behaviors hypothesized for *Homo erectus*—cannibalism.

Cannibalism Is Now Back in Vogue, But What Is the Evidence at Longgushan?

The practice of eating human flesh was known to the ancient Greeks as "anthropophagy" and it was reported by Herodotus. But we know the practice by the much more common name of "cannibalism." "Cannibals" were first named by Christopher Columbus, who brought back several "Carib" Indians from the New World to Spain for exhibition at the royal court. Somehow the "r" in Carib became transmuted to the double "n" in Columbus's transcription of the term, even though the sea that was also named for the tribe was more accurately rendered. The Caribs were a highly mobile and warlike society, attacking Arawak villages on islands throughout the Caribbean. Contemporary chroniclers reported that marauding Carib war parties would cook and eat the hearts of their dead enemies, a practice as fascinating to Europeans as it was repugnant. Many more instances of

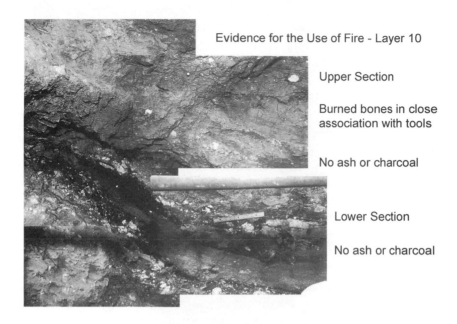

Evidence for the Use of Fire - Layer 10

Upper Section

Burned bones in close association with tools

No ash or charcoal

Lower Section

No ash or charcoal

Evidence of fire at Longgushan. *Top:* A close-up view of the sediments of Layer 10 from the western wall of Locality 1 showing the association of burned bones with stone tools. *Bottom:* Burned mammal bone (probably a fragment of deer rib) derived from Layer 10. It is unclear how this bone relates to hominid activity, but other bony remains from the site suggest intentional roasting of meat by *Homo erectus.*

cannibalism were discovered, or at least reported, in the ensuing four centuries, but rarely were they reported by reliable chroniclers. The reports were so embellished in some cases that they were difficult to believe. Later ethnohistorians questioned many of the accounts, including whether Caribs really did engage in the practice to which they lent their name. Nevertheless, the image of dark-skinned natives with nasal septa pierced by bones, gathered around a human-sized cooking pot brandishing upraised spears became indelibly stamped on the popular mind.

Much less widely known but equally important to our story were archaeological discoveries throughout the first half of the twentieth century that showed clear evidence of past cannibalism in Europe. That these evidences were first found in Europe was more a function of there being a greater density of archaeologists there than anywhere else. But they served to confirm that cannibalism was a worldwide human phenomenon. It was in Europe that Franz Weidenreich first encountered the archaeological evidence of cannibalism. Neandertal bones found at the sites of Ehringsdorf in Germany, Krapina in Croatia, and Monte Circeo in Italy all showed signs of butchery that archaeologists have interpreted as cannibalism. Chief among their proofs was the breakage of bone around the foramen magnum, the aperture at the base of the skull through which the spinal cord passes to the brain. Early people had apparently been "headhunters"— beheading their victims and eating their brains. Ethnography helped out with this scenario by furnishing reliable accounts of contemporary cannibalism and the eating of human brains by tribalists in New Guinea. The practice was discovered in the 1920s to be associated with the neurological disease "kuru," and was banned by the colonial governments in the region.

French archaeologist Henri Brueil argued that the relative absence of limb bones at Longgushan meant that Peking Man had been a head hunter.[12] We now believe that this pattern resulted from hyenas' chewing and destroying hominid limb bones. Weidenreich, as an anatomist, was also acutely aware of the parts of the *Homo erectus* skeleton that were missing. After all, their absence did not allow him to completely describe the anatomy of the species, his avowed goal. Weidenreich at first accepted, and then rejected, Breuil's cannibal argument based on limb bones. He felt on firmer ground with the skull remains. Most of the faces of the skulls were missing, but *all* of the skull bases surrounding the foramina magna were gone. Instead there were large, unevenly cracked holes at the base of the skulls. Weidenreich drew on his knowledge of the European Neandertal record to interpret this consistent pattern of damage as evidence of cannibalism.

We believe that Weidenreich's argument for *Homo erectus* cannibalism based on the broken foramen magnum has been disproved. We have shown in an earlier chapter how this pattern of breakage was indeed caused by

Many *Homo erectus* bones bear the marks of carnivores' chewing on them and were probably the remains of hyenas' meals. Longgushan was home to the giant cave hyena, *Pachycrocuta brevirostris*, a species the size of a modern African lion. During the Pleistocene it lived throughout Eurasia and denned in caves. The species is believed to have actively hunted and scavenged large-bodied species of herbivores. This nicely preserved skull is part of a virtually complete skeleton of *Pachycrocuta* excavated from Longgushan and is on display at the Peking Man Museum, Zhoukoudian, China.

breaking the skull to get at the brain—but by giant hyenas, not the hominids themselves.[13] There are hyena bite marks on the tops of the fossil skulls—evidence of the hyenas' massive jaws—and there are no marks left by stone tools around the skull bases.

However, Weidenreich noticed other evidences of damage on the *Homo erectus* skulls from Longgushan that have until now escaped careful scrutiny. In his 1943 monograph on the *Homo erectus* skull he figured an area of Skull V that he claimed had multiple cut marks from stone tools. We reexamined this area on the first-generation cast of the specimen still preserved at the American Museum of Natural History in New York. To our knowledge no researchers had looked at the Longgushan casts for evidence

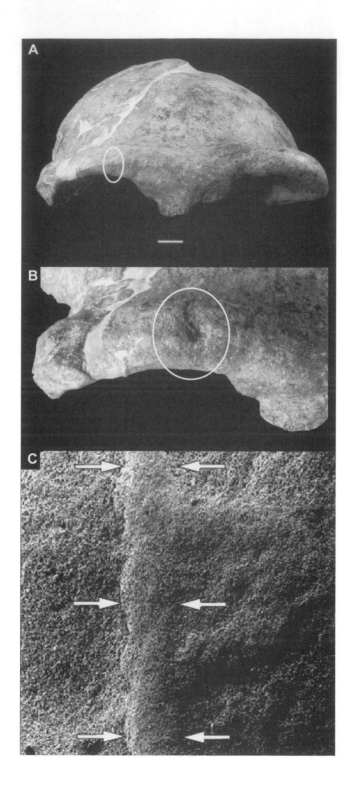

of stone tool cut marks before, and we were surprised at what we found. At low magnification (10×) the surface of the bone is quite well and faithfully rendered by the plaster cast. A number of parallel-oriented, characteristic stone tool cut marks can be seen. The pattern of damage is the same as that produced experimentally by sawing a flake tool back and forth to deflesh bone. We thus agree with Weidenreich that these are stone tool cut marks on Skull V. These cut marks are located on the left skull wall underlying the area of bone to which the temporalis muscle is attached. In the anatomy lab we have used a scalpel many times in this same area to dissect and display this muscle—the largest jaw muscle and the one we use to close our mouths and grind our teeth. We may presume that *Homo erectus*, however, used stone flake tools to remove the temporalis muscle of a dead compatriot, not to study its anatomy or deflesh the body for burial (which we have no evidence for in *Homo erectus*), but to eat it.

In China we examined the surface of the original fossil bone of Skull V, recovered in 1966 and forming part of the same specimen found in the 1936 excavations. We reasoned that if the cast of the temporal bone showed cut marks, then there would be a good chance that the original fossil would show them as well, and in a more reliably preserved context. We discovered evidence of stone tool cut marks on the frontal bone of Skull V, confirming Weidenreich's original observations for this specimen. We believe that Skull V shows that *Homo erectus* cut muscle off the head, and thus very likely engaged in cannibalism at Longgushan.

Modern archaeological analyses at a number of sites around the world—North American Indians, European Neandertals, and many others—now show unambiguously that ancient *Homo sapiens* or their ancestors killed, cooked, and ate members of their own species.[14] The reality of cannibalism is now widely accepted. Ethnographic reports of cannibalism from Polynesia and New Guinea are considered the most reliable, and they indicate a high degree of ritual associated with historical cannibalism, which is imbued with much symbolic meaning. On the other hand there is survival cannibalism, historically well documented and an undisputed reality for people

---

Facing page

Hyena-induced damage on a *Homo erectus* skull. *Top* (A) and *Middle* (B): White ovals depict the location of a probable hyaenid bite mark on the right browridge of Longgushan *Homo erectus* Skull V. This specimen was discovered in 1966 by Chinese excavators. Incredibly, it was found to fit exactly the plaster cast of the back parts of Skull V, which had been discovered in 1934 and 1936 (and then lost during World War II). The frontal bone of Skull V (PA 109) is the only remaining original fossil skull of Peking Man, and thus is useful for detailed study of its surface damage. *Bottom* (C): SEM photograph of an impression of the right browridge of Skull V. The bite mark area shows a shallow groove with a U-shaped cross section characteristic of a large carnivore bite (paired arrows define the groove).

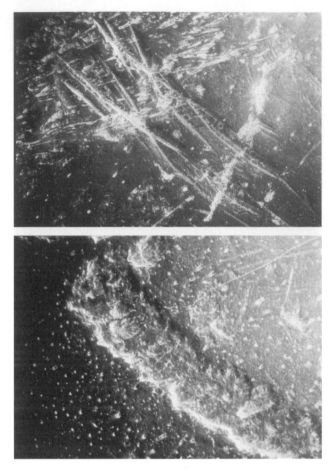

Distinguishing bite marks from carnivores from stone-tool cut marks. *Top:* scanning electron microscope (SEM) photograph of prehistoric cut marks made by a stone tool on
mammal bone from the site of Nihewan in northern China. Stone tools leave bone-thin,
semi-parallel marks with V-shaped cross sections. *Bottom:* Bite mark made by a carnivore
tooth on another mammal bone from the site of Nihewan. Carnivore bite marks are
wider and with U-shaped cross sections. Magnification is about 14×.

in dire circumstances. There is no symbolism here, just the satiation of
extreme hunger. Which, if either, of these possible explanations for *Homo
erectus*'s cannibalism is more likely correct?

Convincing and abundant evidence shows that early hominids as far
back as the late australopithecines 2.5 million years ago used sharp flakes
of stone to cut meat off bone.[15] There is no reason to think that they would
have gone to the trouble if not to eat the meat once it was cut off the
bones. It should also not be too surprising that some of the bones from

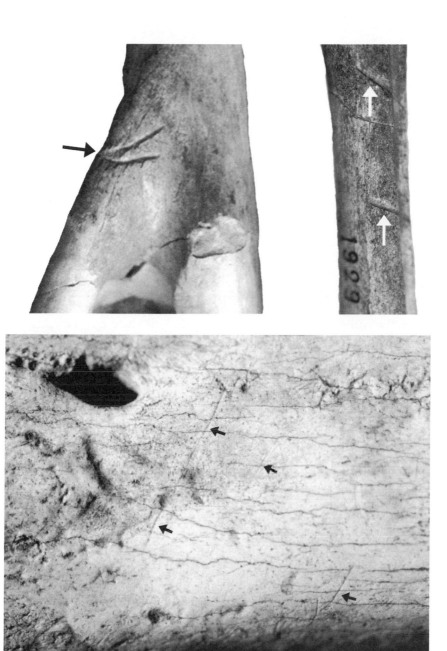

The signature of *Homo erectus. Top:* Bite marks made by carnivores on an upper foreleg bone (humerus) of an ungulate (left) and on a foot bone (metapodial) of another ungulate (right) from Locality 1 at Dragon Bone Hill. *Bottom:* Stone-tool cut marks made by *Homo erectus* on a mammal bone from Locality 1 provide clear evidence that hominids were in the cave at Dragon Bone Hill. Stone-tool marks tend to overlie carnivore bite marks when the two occur together, implying that the large carnivores killed the prey and that *Homo erectus* scavenged it.

which meat was stripped happened to be hominid. Early *Homo* from the Gran Dolina, Spain—dated to approximately 780,000 years ago, somewhat older than Longgushan *Homo erectus*—also shows the telltale cut marks of butchery by his conspecifics.[16]

Speechless Cannibals and Speculations on the Mind of *Homo erectus*

From a variety of investigative approaches, we have concluded in this chapter that two of the most popular, time-honored, and long-held conceptions about early humans—inability to speak and cannibalism—were true of *Homo erectus*. Indeed, the lack of speech was part of the first name proposed for a primitive human, "*Pithecanthropus alalus*" (meaning "ape-man without language"), coined by Ernst Haeckel in 1868.[17] Eugene Dubois borrowed Haeckel's term to name his fossil skullcap and femur from the Javan *Pithecanthropus erectus* 25 years later. As it turns out, Haeckel's hypothetical species name still seems apropos for *Homo erectus*. This speechless hominid, however, could make and use stone tools. Judging from the evidence of stone tool cut marks on *Homo erectus* fossils from Longgushan, we can state that some hominids defleshed other hominids. We deduce that they did this for the same purpose that they cut meat from other animal remains, to eat it. We know from the sequence of cut marks overlying carnivore bite marks at Longgushan and elsewhere that most of this meat was scavenged, having been brought down by the carnivores and not *Homo erectus*.

A human species that can only communicate by grunts or gestures, that scavenges half-rotten kills and eats its own kind, and that habitually bashes others over the head, is one that we might find almost comical. But *Homo erectus* was also a species that periodically unleashed a powerful elemental force, fire, that forced all other species to flee before or yield to it, and one that was resourceful enough to disperse and live successfully in habitats ranging from tropical heat to Ice-Age cold. *Homo erectus*, a successful species that lived a million years, presents a fascinating and unique picture of an alternative humanness. What other deductions can we make about its life?

A *Homo erectus* fossil from Kenya (ER 1808) showed bony evidence of a painful and eventually fatal overdose of vitamin A.[18] This was probably from eating too much carnivore liver. Theorists have noted that a person with hypervitaminosis A takes weeks or months to die, and they suggest that 1808's survival during this time must have meant that members of her group cared for her. If this deduction is correct[19] it is the first evidence

Carnivore and *Homo erectus* damage on the same bone. SEM photograph of an impression of the surface of a fossil jawbone of one of the common deer at Dragon Bone Hill, Locality 1. A large circular impression on the bone records the puncture of a hyena's canine tooth. Immediately above it is the fine, linear incision made by a stone tool wielded by *Homo erectus*. Magnification is about 17×.

for humanlike compassion in the fossil record. Group cohesiveness and cooperation may have been a major element in the species's ability to survive in difficult circumstances.

Eating a carnivore with a liver large enough to give one too much vitamin A, or alternatively, eating an excessive number of livers of small carnivores, might imply successful or systematic hunting by *Homo erectus*. The last generation of paleoanthropologists would have accepted this deduction readily, but today there is a more skeptical climate of opinion surrounding hunting, especially big-game hunting.[20] Abundant archaeological evidence exists to show systematic butchering of large animal carcasses by *Homo erectus*—elephantine creatures like deinotheres in Kenya and mammoths in Spain—but precious little can substantiate the assumption that it was hominids who killed these beasts in the first place. Our early *Homo* ancestors may have been adept at collecting small, relatively defenseless game but incapable of dispatching powerful animals larger than themselves. Brashness, a higher primate character, likely played an important part in displacing ecological competitors and taking their abandoned kills. Fire would have been the ultimate backup to the bluff, and even if the hominids could not summon it as readily as can modern humans, carnivores would need only one bad experience of hominids with fire to learn to be wary.

Cannibalism is a set of behaviors culturally embellished by modern humans. Societies that practice cannibalism have rituals and strong belief systems that govern when and how they eat people. "Human" is not just another item on the menu for *Homo sapiens*. But it may have been for *Homo erectus*. Cut marks on the bones of hominids look just like the cut

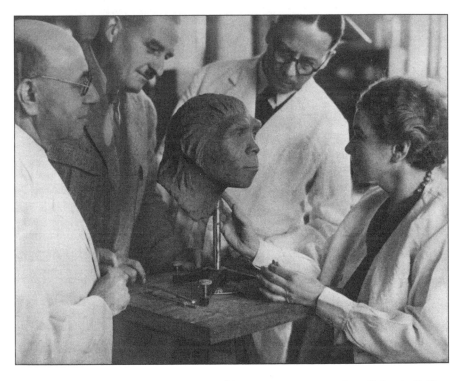

Imagining *Homo erectus*. Lucille Swan (*right*) was an American sculptor living in Beijing in the 1930s. In 1937 she created, under the scientific direction of Franz Weidenreich (*left*), a soft-part reconstruction of Weidenreich's composite skull reconstruction of *Homo erectus*, nicknamed "Nellie" because it was thought to represent a female. Weidenreich first learned of the disappearance of the Peking Man fossils from his Chinese colleagues after World War II, when they sent him a card with the cryptic query, "Where is Nellie?"

marks on the bones of other animals, and hominid bones at Longgushan and other sites are scattered about just the same as other animal bones. *Homo sapiens* usually deal with the remains of cannibalized humans differently than those of animals because human remains carry a symbolism and importance above and beyond their value as food. Evidence of a sense of ritual and symbolism, and evidence of burial of human remains, are first seen among the Neandertals. *Human* attitudes about cannibalism may be what is unusual here; *Homo erectus* cannibalism may be much more in line with what other mammalian meat-eating species do. Most species that habitually eat meat will eat members of their own species if they die.[21] For species that scavenge a significant portion of their food, like pigs, hyaenids, and *Homo erectus*, this behavior may be even more common. There is no evidence or expectation that the *Homo erectus* individual defleshing Skull V at Longgushan recognized the humanness of the meat that he (or she) was collecting.

Imagining how *Homo erectus* may have viewed the world is as fascinating as it is difficult. Archaeologist Thomas Wynn attempted to integrate the types of stone tools made by *Homo erectus* with concepts borrowed from developmental psychology in order to divine something of the "mind of *Homo erectus*."[22] Wynn maintains that earliest *Homo* made stone tools that were conceptually only as sophisticated as those of an ape—sharp, haphazardly formed flakes of rock that could cut. With *Homo erectus* and its signature tool, the bifacial "hand ax," Wynn suggests that cognitive evolution had progressed to a new level. There was now a conceptualization of overall tool shape—symmetry and bulk ("spatial amount")—that *Homo habilis* and apes lack. Wynn believes that the evidence indicates that the mind of *Homo erectus* was able to "construct a more complex external world" and that a *Homo erectus* individual could "coordinate a greater number of and variety of concepts at the same time." We agree that some sort of cognitive advance had occurred in *Homo erectus*. But equally intriguing is the fact that whatever that advance represented, it seems to have stayed virtually static for some one million years, the length of time that bifaces and similar chopping tools dominated *Homo erectus* lithic culture. Perhaps that stasis is related as well to a relative lack of hand-eye coordination in *Homo erectus*, if the small upper spinal cord of the Turkana Boy from Kenya was characteristic of the species. Although many of the details remain blurry, we can agree with Thomas Wynn when he says "*Homo erectus* appears to have been neither ape nor human in a behavioral sense and it is this intermediate status that makes its understanding so important."[23]

Alpha and Omega: Resolving the Ultimate Questions of the Beginnings and Endings of *Homo erectus,* the Species

China has one of the longest recorded histories in the world, a venerable prehistory with many important archaeological sites (in addition to Dragon Bone Hill), and numerous important fossil discoveries that document human evolution through time. China, however, does not seem to preserve evidence of the origins of the hominids.[1] More primitive hominid ancestors, earlier absolute dates for their sites, and molecular-evolution data all point to Africa as the birthplace of the human lineage. It is true that higher primates not ancestral to hominids were present in China and the Far East prior to the appearance of the genus *Homo*, but at least two major population movements out of Africa during the last 20 million years account for these species. One such early population expansion accounts for the presence in China of Old World monkey-apes ("catarrhines") with names like *Dionysopithecus, Laccopithecus*, and *Platodontopithecus*. These small-bodied primates have at times been considered ancestors of modern gibbons or even humans, but recent studies have shown that their anatomy is too primitive for them to be classified as hominoids (the superfamily that includes apes and humans). Their similarities to living species have to be chalked up to parallel evolution. Later expansions of true hominoids out of Africa must have brought to Asia the ancestors of the lesser apes (the gibbon and siamang), as well as the ancestors of *Sivapithecus,* dated in Indo-Pakistan to about 12 million years ago, and possibly ancestral in turn to the Asian great ape, the extant orangutan.[2]

Hominids, our two-legged ("bipedal") variety of hominoid, last shared ancestors with the Eurasian apes about 15 million years ago, perhaps in the form of a species like *Kenyapithecus*. Africa was considered by Charles Darwin the continent on which the evolutionary split of the hominids from the gorilla and then the chimpanzee, our closest living ape relatives, first took place. So far, the fossils documenting this evolutionary divergence have eluded us. A more recent idea is that western Eurasian apes, such as the Greek *Ouranopithecus* or Turkish *Ankarapithecus*, may have been ancestral to African apes, but how they might have traversed the proto-Saharan savannas to arrive in their tropical forest habitats of today remains a mystery. By five to six million years ago, however, hominid ancestors had appeared in Kenya and Chad, and fossil discoveries in Ethiopia, Kenya, Tanzania, and South Africa dating from five to two million years ago document the continuing divergence of the hominids as a uniquely African phenomenon. No fossil evidence of direct hominid ancestors from this time span has been discovered in Eurasia.

From evidence of paleontology and geology it now appears that the wooded and even forested corridors that allowed hominoid populations to move back and forth between Africa and Eurasia were severed beginning in the latter part of the Miocene Epoch. This cutting off of Africa from Asia for a period of some ten million years has a lot to do with the spread of the vast and inhospitable Sahara Desert. Arid conditions over the northern third of the African continent may have been caused by a massive rain shadow extending west from the uplifting Himalayas and by large-scale changes in monsoonal rain patterns.[3] For these reasons, the only hominoids evolving in Asia were apes. There is a conspicuous absence of direct human ancestors in the Asian fossil record in the Miocene and most of the Pliocene Epochs.

Until recently, *Homo erectus* from Dragon Bone Hill was the earliest firm evidence of hominids in mainland Eurasia. Early attempts at interpreting the evolutionary significance of the fossils were hampered by lack of knowledge of the time scale, the geographical extent of the species, and the overall context of anatomical change characterizing human evolution. We are fortunate to have much better answers to these questions today. We now know where *Homo erectus* came from and when it evolved.

African Origins

The first attempt to relate the fossil hominids of Asia to those of Africa was a study by Ralph von Koenigswald and Phillip Tobias,[4] a former student of Raymond Dart at the University of Witwatersrand in South Africa. Von Koenigswald, who after World War II became a professor at the University of Utrecht in the Netherlands, brought to the table his fossils from Java, then assigned to *Homo erectus* and *Homo modjokertensis*. For comparison

Tobias brought from Olduvai Gorge fossils that he, Louis Leakey, and John Napier had recently named *Homo habilis*. They concluded that there were significant anatomical similarities between the two groups of fossils and they postulated a close evolutionary relationship. Their conclusions were prescient but they could not be supported by firm data until later fossil discoveries—some of which have occurred as recently as the last several years—clarified the relationships.

Homo habilis first became much better known through fossils discovered in a region seven hundred miles north of Olduvai Gorge, the Lake Turkana basin of southern Ethiopia and northern Kenya. One of us (N. T. B.) with F. Clark Howell discovered a partial skull with teeth dating to two million years ago in the Omo deposits; we recognized this fossil as the first *Homo habilis* found outside Olduvai Gorge.[5] Richard Leakey and his colleagues discovered more complete skulls of early *Homo* (though most of the first ones were without teeth) in the nearby East Turkana (then Rudolf) deposits. Specialists still disagree on whether all of this assemblage of early *Homo* represents one, two, three, or even four species. Boaz and his colleagues studied this assemblage of early *Homo* and considered the evolution of *Homo habilis* to early *Homo erectus* a good example of the gradual mode of change.[6] Phillip Tobias, who later wrote a monograph on *Homo habilis* rivaling Weidenreich's on *Homo erectus*, has a similar interpretation.[7] The Turkana hominids became the most firmly dated and best-documented assemblage of fossils for the evolutionary transition of *Homo habilis* to *Homo erectus*.[8] At Turkana the first *Homo* fossils are present at 2.4 million years ago, their earliest dated appearance on earth.

In Asia further discoveries of *Homo erectus* were made in Java,[9] and von Koenigswald's old site of Perning, which yielded the Mojokerto skull of early *Homo*,[10] was redated to 1.8 million years ago.[11] The earliest early *Homo* so far found in mainland Asia was dated by Chinese colleagues and one of us (R. L. C.) at the site of Longgupo, China, to 1.9 million years ago.[12] Its full significance is yet unclear since it is only a partial mandible with teeth, but it documents the presence in China of *Homo* before the time of *Homo erectus*. What the Longgupo early *Homo* population very likely looked like was dramatically revealed by discoveries at Dmanisi, Republic of Georgia, three thousand miles to the west.[13] This site has now yielded several mandibles and three partial skulls of early *Homo*, dated to 1.7 million years ago. The skulls are nearly the same age and have very similar anatomy to one of the skulls from Turkana (ER 1813), eight thousand miles to the south. We consider Dmanisi to be *Homo erectus ergaster*, and ER 1813 to be *Homo habilis*. The early African *Homo erectus* has also been assigned by some to *Homo ergaster*, a species name created by anthro-

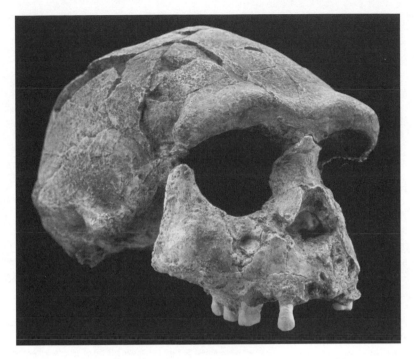

Chinese *Homo erectus* showed close anatomical similarities to, and was closely related to, *Homo erectus* in Java. This is a photograph of the most complete specimen of the species from Java, Sangiran 17. This skull is dated by radiometric methods to 1.25 million years ago, about twice the age of the oldest Longgushan *Homo erectus*.

pologists Colin Groves and V. Mazak in 1975 for an early *Homo* mandible from Kenya (ER 992).[14]

If our reading of the fossil record is correct, then earliest *Homo erectus* (subspecies *ergaster*) can be established as being present in both Africa and Eurasia by approximately 1.7 million years ago. Its immediate ancestor seems to have been *Homo habilis* dating to about 2.4 million years ago and only in Africa. But where did this hominid population come from? We believe that the answer is unequivocally from the African genus *Australopithecus*, discovered by Raymond Dart in South Africa in 1924, just three years after the first teeth of Peking Man had been dug out of Dragon Bone Hill. Whether the ancestral population was *Australopithecus africanus*, Dart's originally proposed species, or another of several more recently discovered australopithecines, remains to be seen. But there is no fossil evidence (even though there are plenty of fossil sites) and little probability of an ancestral source for the genus *Homo* in Eurasia prior to two million years ago. Peking Man and other Eurasian early *Homo* populations ultimately derive from Africa.

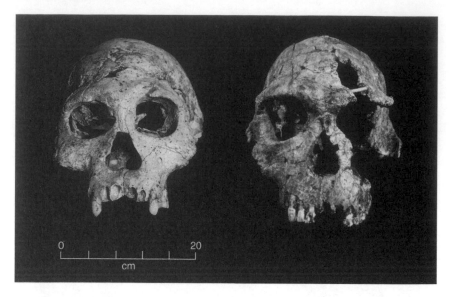

The first *Homo erectus* immigrant to Eurasia. A recently discovered fossil skull (*left*) of *Homo erectus* from the site of Dmanisi, Georgia (specimen number D2700) is the earliest record of the species outside Africa. Dated to approximately 1.7 million years ago, populations similar to those at Dmanisi were likely ancestral to later, Chinese *Homo erectus* at Longgushan. Dmanisi in turn was likely descended from African *Homo habilis*, represented by such fossil specimens as ER 1813 (*right*) from east of Lake Turkana, Kenya.

Africa: The Leaky Crucible

If hominids first arose in Africa, as we believe, the circumstances surrounding their dispersal a couple of million years ago are important to ascertain. Africa had been the crucible of human origins but for some reason it began to leak around two million years ago. We have suggested that the leak first started when expanding dry country in the area of the proto-Sahara Desert began to push hominids before it, to the north and out of Africa.[15]

Until recently it was most reasonable to make sense of *Homo erectus*'s evolution with the following hypothesis: *Homo habilis* (or *Homo ergaster*) gave rise to *Homo erectus* about 1.5 million years ago in Africa, and the species then dispersed to Eurasia. However, with the dating of three hominid sites in different regions of Eurasia (Mojokerto/Sangiran in Java, Dmanisi in Georgia, and Longgupo in China), all significantly earlier than 1.5 million years ago, it became clear that hominids existed outside Africa before this time. Until the fossil cranial remains were discovered at Dmanisi, the identity of this hominid was uncertain. Now it is clear that it was anatomically very similar to African late *Homo habilis* and earliest *Homo erectus*. The initial deductions of Tobias and von Koenigswald about a close African-Eurasian link had proved correct.

With these new data, one of us (R. L. C.) hypothesized that *Homo ergaster* (or *Homo erectus ergaster* as used here) dispersed from Africa and evolved in Asia, becoming a species that we can formally designate with a subspecies name as *Homo erectus erectus*.[16] An equally persuasive case can be made for the evolution of *Homo erectus erectus* in Africa from the same species.[17] We believe that we can accommodate both hypotheses and the observations on which they are based within a new interpretation of hominid evolution.

First of all, if the same evolutionary transition occurred in Africa and Asia—same species transition at the same time—they must be linked. The populations of *Homo erectus ergaster* in Africa 1.9 million years ago must have been in genetic contact with populations of *Homo erectus ergaster* in Asia. This sounds remarkably like Weidenreich's 1947 idea of hominid evolution and its modern version, multiregional evolution—a widely spread-out species with populations connected by gene flow evolving on a wide front. Despite a basic similarity, new genetic data and interpretations argue against the specifics of such a model.

A massive amount of molecular data from living humans shows that the evolutionary branches of biomolecules that we can compare among human populations are short. These particular molecules evolve rapidly and they converge to a common ancestral molecular configuration not older than about two hundred thousand years.[18] Molecular geneticists have tied this biochemical origin of the typical human genome to fossil evidence documenting *Homo sapiens* in Africa at about the same time. Many paleoanthropologists have agreed that these data confirm that *Homo sapiens* appeared at this time in Africa and then spread out into Eurasia. We, however, believe that the molecular data have been widely misinterpreted, both by most molecular geneticists and most paleoanthropologists. The missing perspective comes from an older branch of genetics, one that came to prominence in the first half of the twentieth century—population genetics.

The conception of human expansion out of Africa held by many molecular geneticists, whose primary data we must remember come from laboratory test tubes and electrophoresis gels, is that of an abstract group of hominids surrounding an ancient African female named, for marketing purposes, "Eve," of course. Paleoanthropologists, whose primary data are merely isolated bits of fossil bone and chipped stone, have also conceived of the epic expansion of hominids into Eurasia as something like a single band of scruffy, spear-toting hominids walking into Asia Minor, peering over the horizon into the promised land of Eurasia. With as little data to go on as we apparently have, it is perhaps understandable that such B-grade cinematographic short takes are tolerated. One can almost hear the leader of the band exhorting the group to go forth and multiply.

Population genetics is a field of study that uses mathematical methods and field studies of modern species. Its conclusions can escape the notice of paleoanthropologists, whose research interests focus on the past, and molecular biologists, whose research interests focus on the laboratory. However, calculations in population genetics show that a single band of hominids, and certainly not a single woman nicknamed "Eve," could not have come across from Africa to "replace" all the hominid populations. But neither did regional populations of hominids in Eurasia simply absorb the newcomers from Africa into their gene pool. The residents were replaced as genetic species. It just happened over time and with many populations—a story with perhaps a less dramatic storyline but one that is ultimately more compelling because it is more likely what actually occurred.

A New Evolutionary Model: Clinal Replacement

Population geneticists have demonstrated that species evolve within groups of organisms called populations. Genes in populations do not just float around willy-nilly as they pass from parents to offspring. There are specific rules and regularities between such things as population size and natural selection that determine how the population will actually evolve. If a population shrinks down to a very few breeding individuals, for example, a lot of genetic diversity will be lost, and the population will pass through a so-called bottleneck. Mathematical relationships of population size and genetic diversity have been worked out for many species. Calculations suggest that in a simplistic model of a founding band of hominids to account for all living humans, with the level of molecular change that we see in *Homo sapiens*, world populations would have been only about ten thousand people. This is much too small a number and it strongly suggests that something major is wrong with the population models for human evolution.

Geneticist Elise Eller of the University of Colorado has hypothesized that a pattern of successive population extinctions and recolonizations could explain such a molecular pattern in human evolution.[19] Her results do not support predictions made by the multiregional hypothesis but they do provide a basis for understanding the apparent discrepancy between what must have been large population sizes ("census size") of early *Homo* (that we deduce from the geographic spread of fossils) and the small "effective population size" (that is indicated by the genetic data). Genetic diversity as judged by rapidly evolving biomolecules was lost as local populations went extinct, giving an artificially low estimate of past population size. A neighboring and related population, carrying a number of genes of the extinct population through its clinal genetic connections, then moves into the region and occupies it. If such a pattern of evolutionary change were repeated for thousands, tens of thousands, and then hundreds of thousands of years, we would see the resulting short branches of rapid evolutionary molecular change.

Eller's observations are profoundly important, but they leave us in a quandary: How do we make sense of the new and compelling population genetics perspective, and at the same time account for the apparent synchrony of evolutionary change as seen in the fossils? We need a model that can explain both the relatedness of widespread past hominid populations, as we see at 1.9 million years ago in Africa and Eurasia, and a mechanism to explain the short histories of biomolecules in human ancestry.

We propose a hypothesis that explains both our independent observations on the fossil hominids and many geneticists' observations of the molecular evolutionary data. *Homo erectus ergaster* evolved into *Homo erectus erectus* in both Asia and Africa at the same time because of genetic connections among populations, but data of rapidly evolving biomolecules from modern human populations do not record this relatively ancient event. The rapid mode of molecular evolutionary change in *Homo sapiens* has served to overprint prior genetic change. We believe that this overprinting has conspired with a more complete fossil record for later phases of human evolution to give the impression of total replacement of species, as hypothesized by "Recent African Origin," "Out of Africa," and "Mitochondrial Eve" replacement theorists.[20] Our model, we believe, explains observations of both Replacement and Multiregionalist camps and makes sense of the fossil evidence in a defensible genetic and populational model.

The extant human species exists as clines—geographically defined populations with different gene frequencies (and physical traits) that intergrade at the edges. In the past this broad geographic clustering of physical characteristics and gene frequencies has been termed "race." Race has become an unpopular term because of its long history of association with human rights abuses and genocide (as in Nazi-era "racial purity," U.S. "segregation," South African "apartheid," and Serbo-Croatian "ethnic cleansing," to name just a few examples). But biological anthropologists have recognized and studied geographically circumscribed human variation for a century and a half. It exists. Geneticists measure it and forensic anthropologists can recognize it in the bones of murder victims. "Race" is biological variation with a geographic component. It is frequently confused with but is distinct from "ethnic" variation. An "ethnic group" is a culturally defined unit also originally circumscribed by geography but primarily defined by

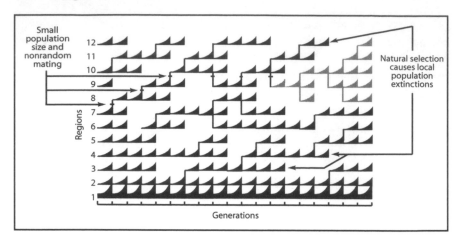

Clinal replacement. An idealized population spread across 12 regions evolving for 19 generations was modeled by population geneticist Sewall Wright (1940). Stylized population-growth curves increase, and then connect forward to a new generation or connect to an adjacent region by gene flow. A cline is a related group of populations within a species that show traits varying along a gradient of geography. If we walked from region 1 to region 12 at generation 9 in this figure, we would traverse a cline. There would be slightly greater genetic discontinuity when crossing from region 5 to 6, and from region 8 to 9, because of decreased gene flow between these regions. At certain times, such as at generation 6 in region 12, a cline goes extinct. Related but distinct populations from adjacent regions then spread into and repopulate the region. It is this mechanism that we hypothesize ultimately accounted for the appearance of new species in human evolution, as when *Homo erectus* evolved from *Homo habilis*, or when *Homo sapiens* evolved from *Homo heidelbergensis*. Clinal replacement explains how such species transitions can show continuity in traits within regions while at the same time accounting for the appearance of new species in evolution.

language, religion, custom, dress, and other learned behaviors. Ethnicity and culture coexist with and exert many important influences on the genetics of populations, but they are not the same thing.

Geneticists today are fond of saying that human races do not exist because the variation found within a population equals or exceeds that found between and among populations. This statement is both profoundly important from the standpoint of understanding gene flow in human populations, and virtually meaningless from the standpoint of whether or not we use the term "race." It is profoundly important because human groups tend to outbreed, that is, members of a group find attractive, mate with, and reproduce with members outside their immediate groups. This tendency to outbreed is termed "exogamy"—"marrying outside." Exogamy is a culturally mediated behavior and it acts to blur the edges of geographically delimited differences in populations. Exogamy is counteracted by geographic distance and physical barriers. The farther apart two people are the less likely they are to get together to have children. Similarly, if mountains, rivers, chasms, deserts, oceans, glaciers, or other physical barriers separate them, they are not likely ever to meet each other. The meaningless part of the above statement comes in when we choose a "population" that is based on political, tribal, or linguistic criteria but that spans large tracts of territory and crosses multiple physical barriers. Of course we will find such a contrived population to be characterized by as much or more biological variation as an adjacent population with less diverse geography and fewer geographic barriers. And we have not even taken into consideration large-scale migration, which greatly confuses patterns of biological variation. But our goal here is to understand how human populations are organized and have evolved, not to quibble about terminology. Let's just say that in general we will find a geographic cluster of biological characteristics in a human population group and that there will usually be a gradient of change of those characteristics into surrounding groups. This is the definition of a "cline."

There are many examples of clines of human genetic and physical traits. A gene that causes dry ear wax, for example, is very common in the Far East, and as one moves westward it becomes progressively less common, until it virtually disappears in populations in the British Isles. The cline may exist because of the exogamy of innumerable small populations over millennia—populations passing the genes along like runners handing off batons in a relay race. Alternatively or additionally, the spread of this gene may have been facilitated by mass population movements east to west, such as the Mongol invasion of the thirteenth century. Physical traits that show clinal variation in humans include skin color, which tends to be light in high latitudes and darker near the equator, and body form, which tends to be slender and linear near the equator and compact and rounded in colder climates. There are many exceptions to these generalities, but they are broadly true. We know that early African *Homo erectus* (the "Turkana Boy") at Nariokotome had the same linear proportions of limbs and trunk that Africans who live there, near the equator, still have.[21] We would predict, but we cannot yet demonstrate, that the hominid populations occupying higher latitudes in Eurasia would have relatively shorter limbs and stockier builds. These bodily characteristics would have changed gradually as one moved away from the equator, as would be predicted by the clinal replacement model.

In discussing genetic change as measured by molecular evolutionists investigating human origins, it is important to remember that they are looking at genes and their protein products that evolve rapidly, that is, that change on the order of every few thousand years. Most of the remainder of our genes evolve quite slowly, and because they do not gauge the time frame for recent events in human evolution, they have not come into the discussion. But in our discussion of clines the majority of the human genome, not just the small, rapidly evolving portion, is important. We must understand that when a clinal neighbor replaces an extinct population, *most* of its genes will be the same. This is quite obvious with a moment's reflection. We share basic characteristics too numerous to list with life forms ranging from the single cell to primates, and all of these physical and physiological traits are controlled by homologous genes. These basic, ground plan genes that control much of our biological formation are not replaced when one clinally related population of a species replaces another. There is continuity in most gene lines and we consequently see this continuity in anatomical traits through time. All too often in debates between molecular-based and fossil-based theorists, a false equation is made between lineages of rapidly evolving genes or biomolecules and lineages of populations of a species. They are in fact very different.

Some of the confusion surrounding the interpretation of molecular and fossil data in human evolution probably originates within genetics itself, spawned by unresolved and divergent viewpoints from biochemical genetics and population genetics. When Rebecca Cann, a fellow graduate student of ours at Berkeley in the 1970s, was developing the "Mitochondrial Eve" hypothesis, she did so in collaboration with Allan Wilson, a professor of biochemistry. Wilson had earlier collaborated with anthropology professor Vince Sarich, himself an anthropologically trained biochemist, to research the ape–human split. The theoretical underpinning of this revolutionary approach to understanding human evolution came from the seminal understanding of the "molecular clock" discovered by biochemists Emile Zuckerkandl and Linus Pauling.[22] Cann, as had Sarich before her, took their internally consistent and tightly argued molecular conclusions and applied them directly to interpretations of the fossil record. Paleoanthropologists, who for the most part do not accept data that they cannot see with the naked eye, were left scrambling for resolution. We believe that the missing perspective is population genetics, first integrated with hominid fossils in the evolutionary hypotheses of Franz Weidenreich via his contact with and citing of the work of geneticist Theodosius Dobzhansky, discussed earlier.

Population genetics in general is undertaken by a very different group of scientists than is molecular genetics. Mathematics plays an important part in the theoretical formulations of population genetics; also, work with living and breeding organisms, such as fruit flies, has traditionally formed a central part of the experimental work in this discipline. Some of the great names of population genetics are Dobzhansky, Sewall Wright, J. B. S. Haldane, Ronald Fisher, and Maynard Smith. None of their works or ideas, however, played a major part in the biochemical genetic formulations of recent Replacement models in human evolution. Yet population genetics occupies a central place in the "synthetic theory of evolution"—an emergent perspective after World War II, and it affected paleoanthropology dramatically through the "new physical anthropology" of Sherwood Washburn.[23] The population perspective of clinal replacement, we suggest, will allow distinctions between which molecules, genes, and physical traits are being compared, and it will facilitate determining the appropriate time frame and geography for discussions of human evolutionary dynamics. Eller's recent paper, mentioned above, is a good start.

The structures of past hominid populations are important to understand if we are to deal with past clinal connections among them. Population density among mammals is related to their body size. As body size increased from *Homo habilis* to *Homo erectus*, population density and home-range size also increased. This happens because larger bodies require more food resources.[24] If the species is an omnivorous or carnivorous one its home range tends to cover a much larger territory than if it is a herbivorous species. As it spreads out to obtain more food, it will come into contact, and probably conflict, with neighboring populations. We believe this model explains much of the overland movements of early *Homo* populations. But once humans were in Eurasia, geographic barriers, such as bodies of water, which affect how and where populations can move, exerted important influences on subsequent evolution.

Dispersal and Evolution of *Homo erectus* in Southeastern Asia

Since early humans did not possess watercraft or the ability to cross large bodies of water, they had to rely on dry land or shallow water to move from one area to the next. A lowered global sea level caused by massive amounts of water being locked up in glaciers produced the corridors needed for human dispersal. Islands in southeastern Asia opened up to early *Homo erectus* as the shallow continental shelf extending from mainland Asia was exposed. Exposure of the so-called Sunda Shelf would have formed "Sundaland," a large extension of the Malay Peninsula that linked the islands of today's Indonesian Archipelago (including the islands of Java and Borneo) with the Southeast Asian mainland. With present ocean floor conditions, a 30-meter drop in sea level would have linked Java with the Southeast Asian mainland.

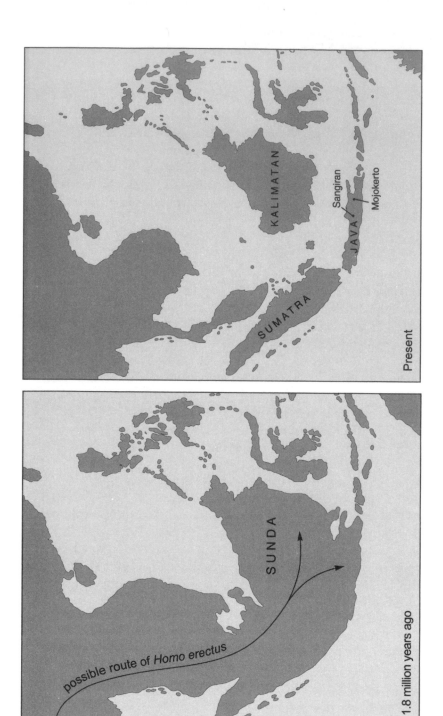

The ancient geographical connection of Java and mainland Asia. *Left:* "Sundaland" or Sunda is the shallow seabed off the coast of mainland Southeast Asia that would have been dry land when lowered sea level exposed it about 1.8 million years ago. Arrows indicate the probable migration route of *Homo erectus* and other terrestrial animals. *Right:* Southeast Asia in modern times. Much of Sunda today lies under the shallow seas of island Southeast Asia.

Java is a part of an extended Southeast Asian peninsula that has a rich fossil record of *Homo erectus*. Sundanese Asian *Homo erectus* localities differ from their African and Eurasian counterparts in three significant respects. Stone tools, one of the hallmarks of *Homo erectus* in Africa and Eurasia were rare in Sundaland. This suggests that the tool assemblage used by Sundan *Homo erectus* probably centered on perishable materials and differed from that used in Africa and Eurasia. Secondly, the large mammal fauna of Sundaland on which *Homo erectus* preyed or scavenged had limited diversity, in stark contrast to the broad variety of large mammals known to have lived with *Homo erectus* in East Africa. Finally, Sundaland, a terrain of extensive seacoasts and muddy shorelines, presented very different habitats and resources from the inland areas occupied by *Homo erectus* in Africa and Eurasia.[25] Central Javan localities such as the Sangiran Dome, Trinil, Kendungbrubus, and Perning (Mojokerto) preserve a range of lowland estuarine, deltaic, and riverine environments. The occurrence of volcanic rocks also points to the presence of nearby volcanic highlands. It is likely that if *Homo erectus* arrived in the area when sea level was 75 meters or more lower than it is today, a large east-flowing river system, the East Sunda River System, would have provided a resource-rich corridor from western Sundaland to the southern coastal region. Such a varied physiography would have sustained a diverse patchwork of plant and small animal communities that presented *Homo erectus* in Java with a range of ecological opportunities.

As the Pleistocene progressed there are indications that Southeast Asian *Homo erectus* became more distinct and isolated from populations to the west and north. They may have intermittently been cut off from global hominid gene flow by sea level changes for about a million years. This isolation may well have fostered the survival of *Homo erectus* or its little-changed descendant populations on Java much longer than in mainland Asia or the rest of the world.[26] In the greater Africa-western Eurasia hominid population, a new species of hominid (*Homo heidelbergensis*) evolved, and this species moved throughout mainland Asia, replacing *Homo erectus* about a half a million years ago. This is the hypothesis of the "Out of Africa"[27] replacement theory as applied to *Homo erectus*. In Java, *Homo erectus* may well have held on much longer, until as late as fifty thousand years ago, when the Ngandong hominids lived.[28]

Climate Change and the Extinction of *Homo erectus*

The fossil evidence, we believe, is now sufficient to show that populations of hominids with species-level anatomical differences replaced *Homo erectus* in Asia. Replacement in mainland Asia happened earlier than in insular Southeast Asia and involved a replacement of *Homo erectus* by *Homo heidelbergensis*. The evolutionary transition in Java was a replacement of *Homo erectus* by *Homo sapiens*, a difference brought about by the relatively isolated population that had evolved without gene flow from mainland populations. This observation forces us to reject the tenets of Multiregionalism and the idea that Asian *Homo erectus* gradually evolved in situ, with only some gene flow from outside.

The abrupt transition to *Homo sapiens* that apparently occurred in Java may be an example of extinction and replacement of populations, and not an example of clinal replacement. Replacement models, like the "Out of Africa" model, differ from clinal replacement in that they hypothesize *total* replacement of the resident hominids. Clinal replacement posits a more gradual population transition than this (but more rapid that Multiregionalism). The fossil evidence may be too sparse to resolve this issue at present, but it is entirely possible that gene continuity may be seen even in this rapid and apparently abrupt transition to *Homo sapiens* in Java.

On the other hand mainland extinction of *Homo erectus* seems to have occurred less abruptly and more in keeping with the hypothesized mode of change in clinal replacement. *Homo heidelbergensis*, an Afro-European, separate species and thus a population that could not produce fertile offspring in matings with the resident population, came in and occupied the entire former range of *Homo erectus*, driving it to extinction. The replacement of *Homo erectus* that we see in the Asian fossil record was by populations of *Homo heidelbergensis* with substantially more genes from the west, not all at once by a separate species. Incrementally, but relatively rapidly in terms of geological time, this process continued until the geographically defined and anatomically discrete genetic constellation that was Asian *Homo erectus* ceased to be. This scenario is fully in keeping with the fossil evidence for replacement, and explains the continuity in traits long cited by Multiregionalists. What further supports it, in our opinion, is the genetic evidence. Total replacement does not satisfy the expectations of population genetics. It cannot resolve the disparity between census size (the global population of hominids), which had to be large, and effective population size, which that model requires to be small. Clinal replacement does resolve this problem by its many small populations moving and replacing other populations.

The model of clinal replacement is in a sense a process of "microevolution," that is, it is a process of small-scale changes in populations. One of the greatest challenges in all of evolutionary biology is to relate phenomena at one level to observations at other, higher levels of organization. In the case of the endemic evolution of *Homo erectus* in Southeast Asia and its subsequent replacement, there had to be a speeding up of the microevolutionary changes to account for the macroevolutionary change that we see in the fossil record. To be a convincing explanation, clinal replacement has to explain this acceleration of genetic and anatomic change.

A species may *evolve* to become extinct, thereby passing on its genes largely intact, but altered sufficiently for its descendants to present a significantly different adaptation. Evolutionary biologists call this "anagenesis." *Homo erectus* over part of its range as a species became extinct because it evolved by anagenesis into a descendant species, *Homo heidelbergensis.*[29]

For the ancestors of *Homo erectus* to move out of Africa, there needed to be both an open geographic pathway of dispersal and a motivating impetus. We discussed these in the context of environmental change that not only opened up "savanna-like" environments but also pushed populations up and out of Africa because of spreading aridity. Paleoclimatic evidence accumulated over the past 25 years has shown that the period just before the beginning of the Pleistocene, some two million years ago, was a time of such climatic change in the Old World. It also corresponds to the dates of the appearance of the first hominids, migrating from Africa, in Asia and easternmost Europe. A "paleoclimatic pump" of encroaching zones of aridity was likely the push out of Africa that impelled the movements of populations.[30]

In order to test expectations of the clinal-replacement model for the extinction of *Homo erectus*, we must now look at the paleoclimatic record in Eurasia over the subsequent million and a half years. What were the events in the real world that might have driven the evolutionary changes that we have hypothesized?

The climatic history of China in the Pleistocene shows a repeating pattern of ups and downs. The oxygen-isotope curve from drill cores in the deep sea that records a continuous global temperature change over the past 1.5 million years shows a sawtooth pattern of fluctuations. These fluctuations come about because heavier oxygen-18 isotopes become preferentially trapped in ice, leaving the lighter oxygen-16 isotopes to become relatively more abundant in the environment. Thus, the ratio of oxygen-18 to oxygen-16 at any particular time in the past provides a record of relative amounts of global ice volume during that time. Ice volume shows a close relationship with global temperature. In China, the deposits of loess settle from the air and become consolidated by rainwater. Their thickness and grain size have been found to match the deep-sea core closely, and both increase during times of cold. Dutch researcher D. Heslop and his colleagues showed that loess thicknesses and grain sizes in northern China record the relative force of cold and dry winter monsoonal winds blowing off the Tibetan plateau during glacial periods.[31] Measurements of magnetic

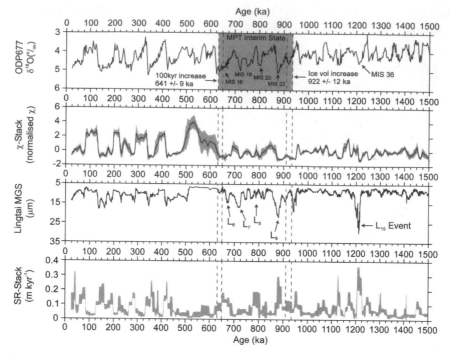

The top of the figure illustrates the oxygen isotope record from the deep sea for the past 1.5 million years, showing significant fluctuations in global ice volume (peaks of the graph represent periods of large ice volume in polar ice and in glaciers, primarily in the northern hemisphere). The next figure records relative strengths of the summer monsoons that dump rain on northern China, as measured by the relative amounts of magnetic activity in iron-containing minerals formed by rain-related soil building. The lower two figures record activity of the winter monsoons, which bring cool and dry continental air from the west. In the figure next to the bottom, larger windblown sediment particles (MGS or "mean grain size") indicate stronger winter monsoons (depressions in the curve), and thus colder and drier conditions in northern China. In the bottom figure, the sedimentation rate shows peaks during glacial times when the winter monsoons blow more loess into northern China and thus create thicker deposits. Abbreviations used in this figure: MIS = marine isotope stage; MPT = Mid-Pleistocene Transition. This model is based on research by D. Heslop and colleagues (2002).

iron-containing elements found in weathered loess corresponded to increased rainfall and periods of increase in force of the summer monsoons blowing in from the Pacific Ocean.

From paleoclimatic studies undertaken at Longgushan and already discussed in chapter 6, we have evidence that the large ungulates, as well as *Homo erectus*, disappeared from Dragon Bone Hill and went south during the harsh Pleistocene glacial periods. To understand how widespread this pattern of climate change and migration was during the Pleistocene, we can compare the period preserved at Longgushan with the overall patterns of climatic change discovered by Heslop and his colleagues. We see that the period of time during which *Homo erectus* is recorded in the Longgushan Locality 1 sediments—670,000 to 410,000 years ago—was relatively warm in its first half, with increasing warm summer monsoons and sustained low-force cold winter monsoons. Beginning five hundred thousand years ago the winter monsoons increased in force, but even so they did not have a major effect on the amount of windblown sediment until the very end of Longgushan deposition, just before four hundred thousand years ago. After this peak in cold, dry, glacial conditions, *Homo erectus* disappears from the record of mainland Asia. The species may have held on in the relatively warm regions of Southeast Asia for some time after this. The Ngandong (or Solo) fossils seem to document such a late-surviving insular population of *Homo erectus*.[32]

The global isotopic paleotemperature curve and the loess records from China demonstrate that there were major shifts between cold and warm periods roughly every one hundred thousand years after the middle Pleistocene. Between 922,000 and 641,000 years ago fluctuations from cold to warm, when they occurred, were greater than before. This turning point in climatic history has been termed the Mid-Pleistocene Transition. Each of these cycles witnessed a major change in the Asian monsoonal pattern, which changed rainfall, vegetation, and animal life. We would predict that during these periods of more significant climatic change clinal replacement would be of greater magnitude and more observable as species replacements.

Geneticist Alan Templeton recently posited an "out of Africa" migrational event that he estimates from molecular clock data to have occurred between 840,000 and 420,000 years ago.[33] There is approximate agreement of these genetic changes with the timing of increased amplitude of climatic change in the middle Pleistocene. Perhaps even more significant from the standpoint of human evolution is that the even greater amplitude of change—colder glacials and warmer interglacials—typified the late Pleistocene and began immediately after the age of the fossil deposits at Dragon Bone Hill Locality 1.

It may have been the increasing cold at higher latitudes and the increasing aridity in the tropics as the Pleistocene progressed that caused the extinction of *Homo erectus* and that drove the evolution of modern human populations. But it may also have been simply that the pace of environmental change was too fast and the amplitude of change was too much for human biological evolution to handle. *Homo erectus* was probably the last species in our lineage to adapt to environmental change primarily by biological means, and that mode of adaptation proved inadequate for the rigors of the Pleistocene.

Anatomy and the Demise of *Homo erectus*

We have presented a hypothetical model that explains the mechanisms of the evolution and eventual extinction of *Homo erectus*. We believe the model is consistent with fossil and genetic evidence. But the evolutionary changes in anatomy, function, and overall adaptation from *Homo erectus* to *Homo sapiens* remain to be explained. Many of the changes that we know of have to do with the head.

Theoretically, a species could have both a commodious skull to house an enlarged brain and a heavily armored and thick skull for protection. Reality steps in when the weight of such a structure has to be supported and balanced atop the spine. Many large-bodied species, whose skulls must be correspondingly large, have bone that is honeycombed inside to reduce its weight. A section through the skull of an elephant or a giraffe is surprising because so much of the inside of it is air, surrounded by paper-thin bone organized into structures called "diploe." Evolving *Homo erectus* had a similar problem with the weight of the skull. If the brain size was increasing, the enclosing bone would also have to increase, but skull weight would have to be minimized. As *Homo erectus* evolved to *Homo heidelbergensis*, skull weight was decreased by lessening the thickness of cranial bone. And, as was discussed earlier, there may have been a "fourth function" that helped account for the decrease in skull thickness—cooling of the brain via more efficient venous blood flow between the skin and the cranial vault.

The defensive functions of the pachyostotic *Homo erectus* skull were lost as *Homo sapiens* evolved a larger, more globular, and thin-walled skull. Human intraspecific violence by no means ended, but other means to protect themselves from trauma or to avoid attack, or both, were evolved by the descendants of *Homo erectus*. Almost certainly, these adaptations were cultural, the hallmark of *Homo sapiens*, and no longer biological.

CHAPTER 9

Testing the New Hypotheses

Dragon Bone Hill and its hominids represent the starting point for many of the major paleoanthropological debates over the past 75 years. The site's discoveries have been at the center of the hypotheses on the origins of the use of fire, the beginnings of human language, the evolution of the brain, hunting, cannibalism, stone and bone tool use, and ancient human diet. The hominids themselves have occupied center stage for most of the time since their discovery, and paleoanthropologists have continued to place them squarely on the direct lineage of human evolution, that is, the lineage leading to *Homo sapiens*. This has been the case despite the fact that *Homo erectus* was endowed with some unusual and unhumanlike anatomical traits. We have attempted to piece together in this volume the anatomical, archaeological, geological, paleontological, and paleoecological evidence to present a series of new hypotheses about this intriguing species, particularly Chinese *Homo erectus* and its main site of discovery, Dragon Bone Hill. We believe that this new composite view fits the available evidence, but all hypotheses in science should be testable. This means that they must be falsifiable—that they can be proved wrong. In this chapter we propose a number of tests which can either disprove our hypotheses or further support them.

Reexamining the Origins of Asian *Homo erectus*

We have postulated that *Homo erectus* in China was descended from a recent immigrant from Africa. Evidence now suggests that the evolutionary transition from *Homo habilis* to *Homo erectus* happened along a very broad evolutionary front—from Africa through Eurasia. More fossil evidence from well-dated sites earlier than Longgushan Locality 1 is needed to test this hypothesis. Our hypothesis is at variance with the view among some Chinese paleoanthropologists that *Homo erectus* evolved in situ in the Far East and is the product of wholly regional differentiation. Fossil evidence from China from between two and three million years ago should help settle the question.

Dragon Bone Hill itself may yet provide evidence of the precursors of late *Homo erectus*. Hominid fossils have not yet been found in the earliest fossiliferous layers (Layers 11 to 17) at Longgushan. Our hypothesis is that this absence of hominids at the near one-million-year mark in the cave reflects not the possibility that hominids were not in China at that point, but that the habitat and environment of deposition in the cave were not suitable for hominids at that time. Very little is known of the earliest levels at Longgushan. One pit ("the Lower Cave") has been excavated and found to have hominids down to Layer 10, and that was in 1929. We predict that hominids will not be found in the lowest levels in the same circumstances of burial as higher in the deposits because the cave was too wet and near river level. Land vertebrate fossils may be found washed in and covered by water-laid sediments, but most of the preserved fauna in these layers will likely be aquatic. Excavation of the lowest levels at Longgushan, coupled with geological and geochemical investigations, could be undertaken to test this idea.

The dates that we accept for the *Homo erectus* fossils from Longgushan are 410,000 to 670,000 years ago, ages that make the most sense from the recent and most rigorous uranium-series, electron-spin resonance, and paleomagnetic analyses, as well as from the chronology of paleoenvironmental isotope curves and loess sediment records. These dates are further back than the traditionally accepted dates for Longgushan and they cover a longer time span. The new dates are less firm than at many sites,

---

Facing page

At two million years ago hominids were restricted to Africa. But by 1.8 million years ago *Homo erectus* had reached Java (a child's skull was found at Mojokerto) and by 1.7 million years ago the species is recorded as having reached western Asia at Georgia (three skulls found at Dmanisi). This map of the Old World shows the location of major hominid fossil sites discussed in this book. This is not a comprehensive list of all sites but rather is a roadmap to the key hominid sites discussed in the text. Areas in white around continental margins indicate land masses that are submerged today but that in the past, when sea levels were lower, provided land links for hominid migrations. The location of each hominid site is indicated by a small black dot. After (or before) each site name is a symbol indicating the different hominid species that can be found at the site.

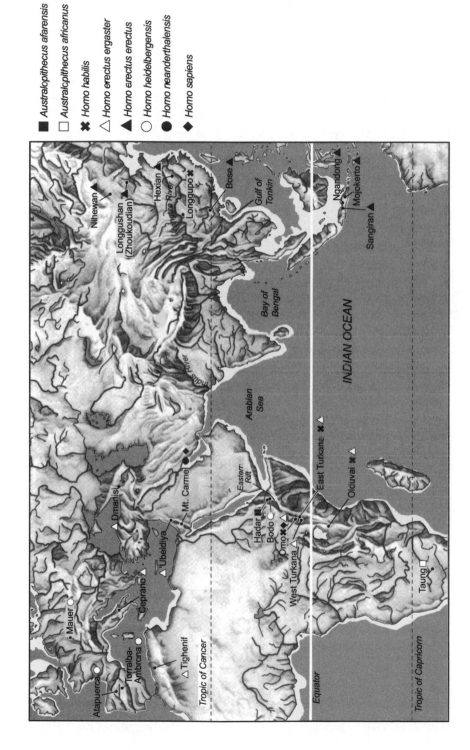

- ■ *Australopithecus afarensis*
- □ *Australcpithecus africanus*
- ✖ *Homo habilis*
- △ *Homo erectus ergaster*
- ▲ *Homo erectus erectus*
- ○ *Homo heidelbergensis*
- ● *Homo neanderthalensis*
- ◆ *Homo sapiens*

Atapuerca
Mauer
Torralba-Ambrona
Tighenif
Ceprano
Ubeidiya
Dmanisi
Mt. Carmel
Hadar
Bodo
Omo
West Turkana
East Turkana
Olduvai
Taung

Nihewan
LongguShan (Zhoukoudian)
Yangtze River
Hexian
Longgupo
Bose
Gulf of Tonkin
Ngandong
Mojokerto
Sangiran

Indus River
Bay of Bengal
Arabian Sea
INDIAN OCEAN
Eastern Rift

Tropic of Cancer
Equator
Tropic of Capricorn

particularly in Africa, that have good potassium-argon dates, and therefore more supporting dating data from Longgushan are needed. Models of human evolution rely on firm chronological frameworks, and valid interpretation of the uniquely important hominid fossil samples from Longgushan require that we subject our dates to continuing scrutiny. Ongoing research in geochronology should be tied to any geological and geochemical research at the site.

The origins of *Homo erectus* now seem to lie earlier in time than the sediments at Longgushan Locality 1 can address. Other sites, perhaps on Dragon Bone Hill itself or in other parts of China, must give us knowledge of the period one to two million years ago, when we hypothesize that *Homo* dispersed and evolved in China. So far only three sites in southern China—Longgupo at 1.9 million years old, Yuanmo at 1.7 million years old, and Lantian at 1.1 million years old—have given us fragmentary tooth and jaw fossils documenting the earliest immigrant hominids. Other sites and sediments need to be found and explored. Exciting possibilities abound in the Nihewan Basin of northern China, which recently yielded artifacts dated at 1.36 million years ago.[1] It is entirely possible that other localities on Dragon Bone Hill that are already known but still poorly investigated could be of the correct age to probe this time period and may hold some important answers. Fruits of this exploration can be expected to shed light on the model of human evolution that we propose—clinal replacement.

Clinal replacement predicts a predominantly gradual mode of evolutionary change from one species to another in human evolution. Replacement of populations occurs, but the replacements are small in scale and involve closely related populations. As we track the incursion of the first hominid species into Asia, *Homo erectus ergaster*, and its subsequent evolutionary change, it will become apparent whether or not this model correctly fits the facts. We and several of our colleagues argued more than 20 years ago that the fossil and geochronological evidence supported a gradual evolutionary transition from *Homo habilis* to *Homo erectus* in Africa.[2] We believe that this argument still holds true and that fossil discoveries made since have confirmed our conclusions. If clinal replacement is correct, a similar gradual evolutionary transition will be discovered between *Homo erectus ergaster* and *Homo erectus erectus* in Eurasia. If an abrupt, species-level change from *Homo erectus ergaster* to *Homo erectus erectus* is found to have occurred in Eurasia, then clinal replacement as a model for Asian human evolution at this stage will be disproved.

Once hominid populations became established in Eurasia, evolutionary change then had the potential to become a two-way street. Genes could flow back into Africa from Eurasia even though the corridor of connection through the Middle East was narrow. Further paleontological investigation of sites accurately dated along the corridor of exchange and throughout

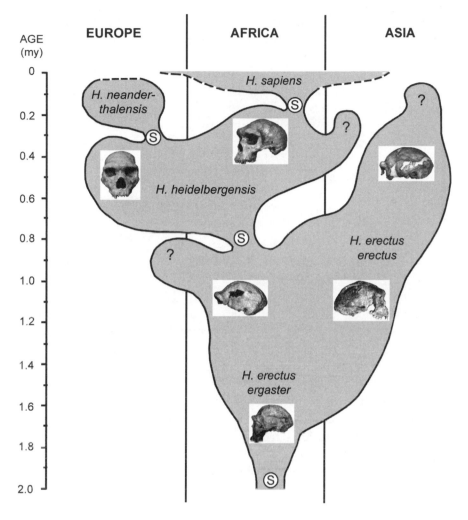

A view of the evolutionary, geographic, and temporal relationships of *Homo erectus*, following Philip Rightmire (2001). "S" refers to speciation events and "my" stands for million years ago. *Homo erectus* persisted in Asia much longer than in other parts of the world, eventually being displaced by either *Homo heidelbergensis* or its descendant *Homo sapiens*.

the entire range of hominid occupation would be essential to testing this idea. Such sites in the one-to-two-million-year-old time range that are already known include Ubediya in the Jordan Valley, Tighenif (Ternifine) in Algeria, Dmanisi in Georgia, Ceprano in Italy, and a number of archaeological sites on the Indian subcontinent that until now have yielded only stone tools. How and when these gene exchanges occurred are important questions to answer. We will continue to use morphological proxies for assessing gene exchange among ancient populations, but we can also hope for the recovery of ancient DNA, already known for Neandertals,[3] to assess these deductions directly. The history of environmental change can also aid us in reconstructing scenarios and determining what actually happened in the evolutionary dynamics of Pleistocene hominid populations.

Charting the Waves and Currents in the Evolutionary Sea

Genes have metaphorically been referred to as existing in "pools" within populations. But for populations that spread widely over continents, as did early *Homo*, perhaps an image of "gene sea" is more appropriate. The forces that cause the currents, waves, and eddies of change are of interest to us because they will explain the course of change and the anomalies in the fossil record that we encounter. But how can we chart them? Deducing patterns of continuous populational change through recourse to the discontinuous data of the fossil record will always be difficult, but we do not think it impossible. Fossil sites represent discrete data points, geographically and temporally circumscribed, that must be connected one to another in a continuous matrix of time and paleoenvironmental change. As this matrix becomes better known for hominids, we may anticipate finding instances of fast gradualism and steep clinal gradients ("tachytely" to use George Gaylord Simpson's term[4]) and slow gradualism and low clinal gradients ("bradytely").

We and other researchers have used a method of paleoecological reconstruction of Europe to posit the changes in past climate and geography that would have isolated populations of Neandertals and then facilitated the later movement of the modern *Homo sapiens* that replaced them into Europe.[5] Our thinking was that understanding a relatively recent transition in human evolution—roughly 150,000 to 30,000 years ago—would help us to understand the many more distant transitions that have occurred throughout the last two million years of the history of the genus *Homo*. We assembled all known records of past climate, including fossil pollen and animal fossils, geological records of maximum extensions of glaciers during the

Ice Age, and the locations of past bodies of water. With these data we constructed a paleo-map of Europe with vegetational zones and major physical features. Onto that template we placed all the known fossil hominid sites of the time. The composite map showed that Neandertals (considered either as a separate species, *Homo neanderthalensis,* or a subspecies, *Homo sapiens neanderthalensis*) occupied a region encompassing Europe and the Middle East (exclusive of North Africa) that was largely cut off from the rest of the world by mountains, glaciers, meltwater rivers, and seas.

Fossils discovered in Europe since this research was done have shown that the origin of the Neandertals from more archaic populations of *Homo,* termed *Homo heidelbergensis,* is still interpreted most conservatively as gradual and progressive in nature. The spectacular finds of early Neandertal-like skeletons at the Spanish site of Atapuerca near Burgos are the most dramatic examples of recent discoveries.[6] Neandertals were peoples who lived in glacial Europe and surrounding areas. They were descended from populations of *Homo heidelbergensis* that had become relatively isolated from the global human cline, becoming a more differentiated population. Their limb proportions, their facial structure (protruding mid-facial regions with large noses), and their advanced culture all suggest a successful adaptation to a cold climate. We can understand how the Neandertals gradually diverged from their predecessors by an evolutionary process of clinal replacement.

Another suite of recent discoveries, however, this time at one of the most venerable of Neandertal sites, Mount Carmel in Israel, is more of a challenge to our clinal replacement model. The fossil humans from this site were studied by Theodore McCown of the University of California at Berkeley and British anatomist Sir Arthur Keith in a landmark monograph published in 1939.[7] Two caves in Mount Carmel yielded fossils that McCown and Keith assigned to anatomically modern *Homo sapiens* (from the cave of Skhul) and to Neandertals (from the cave of Tabun). In the age before radiocarbon and potassium–argon dating, they speculated that Tabun was older than Skhul. Because of the anatomical similarities, they postulated an ancestor–descendant relationship of Neandertal to modern *Homo sapiens.* Absolute dates from fossil bone at Mount Carmel, however, showed that some Neandertals from Mount Carmel were actually *later* in time than anatomically modern fossils, and that some anatomically modern *Homo sapiens* fossils dated to *earlier* levels than Neandertals. Archaeologist Ofer Bar-Yosef of Harvard University and his colleagues concluded that this area had been a region of give-and-take population replacements over a period of some seventy thousand years.[8] What caused these waves of human evolution lapping back and forth on the shores of ancient Israel is important to ascertain because whatever it was, it will likely be a key element to understanding the dynamics of human evolutionary change on a larger scale.

The Pleistocene Epoch, encompassing the extensions of northern hemisphere glaciation known as the "Ice Age," was a time of increasingly severe fluctuations of climate. Climates, as we have seen, did at times become colder and drier (during "glacial" times) but they also became warmer and wetter (during "interglacial" times). These fluctuations in climate may ultimately be traced to variations in the amount of sunlight hitting the earth, the so-called Milankovich cycles, driven by solar system dynamics. Closer to home, changes in Antarctic ice volume, changes in global ocean temperature, and changes in the monsoonal rain patterns in Africa and Asia have been shown to have exerted important effects on animal and plant communities and their evolution. The pattern of increasingly wide shifts in climate from cold to warm began about 2.5 million years ago, the date of onset of northern hemisphere glaciations, and became more and more noticeable as time passed. Local responses to the global pattern of temperature change may vary, however, and this is why paleoenvironmental research is so important at each individual site. As we saw in the case of northern China, local effects that were not expected or accurately predicted by the global climate pattern were importantly mediated by the winter monsoons with their cold, dry blasts of loess-laden wind. Continuing field and laboratory research is needed at all hominid sites to assess the local effects of climate change on evolutionary events.

The stark environmental challenges of the Pleistocene have long been considered the icy anvil on which the human species was forged.[9] We believe that the redated sequence of Pleistocene human populations at Mount Carmel and elsewhere in Israel provides detail to the metaphor. When climates became colder, a larger habitable region was opened to the cold-adapted Neandertals, who spread into the new areas. Although some have suggested clinal gradation of Neandertals to modern *Homo sapiens* at some places (such as at Vindija Cave, Croatia[10]), or even interbreeding of anatomically modern *Homo sapiens* and Neandertals (at Lagar Velho, Portugal[11]), Mount Carmel does not seem to be such a place. When times were right, Neandertals flooded the region, displacing modern *Homo sapiens* to the south and east. If there was any interbreeding it is not apparent in the bones that they left behind. A few millennia later, however, when the climate had swung back to warm and the sweaty Neandertals at Mount Carmel began eyeing cool northern climes with envy, anatomically modern *Homo sapiens* showed up again in the Levant. The Neandertals either retreated to the north or died out locally. By forty thousand years ago they were gone for good. We believe that paleoanthropological reconstruction at this level of detail, supported by in-depth paleontological and geological data, provides a model on which research should be pursued at other sites. A number of such investigations will ultimately test the broader hypotheses, such as clinal replacement. Mt. Carmel is one of the anvils of the Pleistocene—a place where the hammer blows of Pleistocene climatic change formed the modern human species. In such places we might expect to see a more rapid clinal replacement, an "edge effect" between populations.

A well-documented Late Pleistocene scenario of evolutionary transition from one early human population to another in one region, such as the Levant, is informative for interpreting earlier phases of human evolution, ones for which we have less complete information. There are some important similarities between the Neandertals and East Asian *Homo erectus*, for example. Both can be shown to have evolved along a trajectory of gradual change from earlier widespread populations—*Homo erectus ergaster* in the case of Chinese *Homo erectus erectus*, and *Homo heidelbergensis* in the case of the Neandertals. Both lived in geographically circumscribed areas—Asia, south and east of the steppes of the Tibetan Plateau; the Gobi Desert; and the Himalayas, for East Asian *Homo erectus*, and an area bounded by the Mediterranean, northern glaciers, and the Ural and Caucasus Mountains on the east for Neandertals. Both showed regionally characteristic archaeological traits—a lack of emphasis on Acheulean choppers in sites east of the "Movius Line" for Asian *Homo erectus*, and technologies emphasizing large, Levallois-type, spear blades for Neandertals. Finally, both diverged and survived in their core regions while more modern populations, which were later to replace them, evolved elsewhere.

It is possible to explain these data and the respective appearances and extinctions of *Homo erectus* and Neandertals as species transitions, and a number of paleoanthropologists do just that.[12] *Homo erectus* comes from *Homo habilis/ergaster* and is replaced by *Homo sapiens* in the former case; in the latter case, *Homo heidelbergensis* evolves to *Homo neanderthalensis*, which is later replaced by *Homo sapiens*. Such explanations are neat, but they do not explain some critical data. We believe that fossil evidence showing anatomical gradations at the edges of populations, regional clustering of traits that would seem to cross "species" boundaries, and possible evidence of interbreeding between Neandertals, as well as population genetic evidence that shows a large degree of population relatedness within the human species, are not compatible with this explanation. We accept the fossil evidence for a much more abrupt and rapid transition from Neandertals to modern *Homo sapiens*, but our interpretative model of clinal replacement holds this as a likely example of tachytely, and therefore of restricted temporal span and more difficult to document clearly in the fossil record.

We predict that as the fossil record of hominids improves, more and more blurring of the presumed species boundaries will occur, making much more of a "lumped" pattern of species names for Pleistocene hominids. The model of clinal replacement can account for the gradual origins of *Homo erectus* and Neandertals—through the process of small-scale extinctions of local populations, and replacements by related neighboring populations. It is during the extinctions of these populations at "anvil sites," where change seems to be more abrupt, that some disagreement may arise over the mode of evolutionary change. When geography and climate acted to separate populations over time, as happened in the cases of Asian *Homo erectus* and the Neandertals, a more apparent boundary between populations existed. If clines in such cases were of greater magnitudes, further evidence should demonstrate the existence of blurred population edges, and thus clinal, not species, boundaries.

Any one site, such as Longgushan, is not sufficient to test a large-scale hypothesis like clinal replacement completely. However, if renewed research incorporates an intensive multidisciplinary investigation of sediments from all time periods preserved at Dragon Bone Hill, and of indicators of the overall paleoenvironments in northern China during this time span, one excellent set of data can be compared with the predictions of our model.

Recent archaeological research has suggested that there were three waves of premodern human migrations of hominids out of Africa—at 1.6–1.7 million years ago, documented by large core-chopper stone tools; at circa 1.4 million years ago, documented by Acheulean bifacial hand axes; and at four hundred thousand years ago, characterized by advanced Acheulean flaked cleavers.[13] The paleoenvironmental scenario that we have outlined, and the model of clinal replacement, would suggest that there were many more. Furthermore, we would expect that there would have been gene flow back into Africa. Future research will need to investigate these interesting possibilities and accurately date them in order to test our hypothesis.

Examining the Lifeways of *Homo erectus*

We have suggested an unusual behavioral adaptation to explain the unusual pachyostotic cranial anatomy of *Homo erectus*. We hypothesize that head bashing was so common and ingrained a behavior in this species that it led to multiple changes in skull form. This is a hypothesis, first advanced by Peter Brown,[14] that needs much further testing. Taphonomic investigations of other *Homo erectus* crania should be undertaken to see if depressed fractures can be seen in their skulls also. Other anatomical researchers should contemplate and propose tests of whether some other biological function of bone that we have not thought of might have been implicated in creating such a thickheaded species. Archaeologists should investigate this model of *Homo erectus*'s behavior to see if it is supported by their data. And there are interesting questions surrounding the intersection of our "protective pachyostosis" hypothesis and Dean Falk's "radiator brain" hypothesis. How did a thickened skull allow for cooling of an expanding brain in *Homo erectus*, especially since the species also evolved in the hot tropics? Did a requirement for cooling in the very enlarged brain of *Homo sapiens* explain the eventual loss of this suite of pachyostotic features in the skull? Major events must have occurred to make *Homo erectus*'s skull anatomy evolve into *Homo sapiens*'s skull anatomy, but as yet we do not have a clear understanding of what they were.

If our postulated functional interpretation of *Homo erectus* cranial anatomy is correct, there are some profound implications for human behavior because, after all, *Homo erectus* is ancestral to our species through *Homo heidelbergensis*. We have suggested that interpersonal violence was so prevalent in this species for so long (some one and a half million years) that major skull reorganization resulted. The effect of this extended period of evolution on our modern psyche must have been significant. If our model is correct, it means that *Homo erectus* was much more of a "bloodthirsty killer" than ever were the australopithecines, characterized as such by their discoverer and other writers.[15] If we recollect how slowly *Homo erectus* culture changed and how formulaic and unvarying it was for so long, it is chilling to contemplate how mindless, methodical, unquestioning, and unmerciful killing of individuals outside of one's own small group would have been. Is it only coincidence that the same terms spring to mind to describe any of our modern-day genocides, from Bosnia to Rwanda to East Timor? We predict that future *Homo erectus* sites will hold evidence, even more direct than the healed depressed fractures of the Longgushan skulls, of past intraspecies conflict and aggression. Aggression between groups was likely a consequence of the constant expansion and contraction of population ranges and territories, mediated by Pleistocene climate change, that created the ecological setting in which these behaviors evolved in *Homo erectus*. It remains for modern *Homo sapiens* to recognize this ancient behavioral trait as a remnant of a successful, if brutal, Pleistocene heritage, and to modify it as it occurs in the modern world by exercising those other traits that most set our species apart—culture and intelligence.

Homo erectus was not only a killer, however. The evidence from ER 1808, the *Homo erectus* from Kenya with presumed hypervitaminosis A, though sparse, is direct evidence that *Homo erectus* people cared for their own. Even without this evidence, however, we could reasonably infer a high degree of solidarity in *Homo erectus* groups simply because of the environmental challenges that they encountered and must have overcome to survive. Again, if our conception of *Homo erectus* culture is correct, we might characterize their group solidarity as unswervingly loyal, total, and lifelong. There must have been innumerable cases of extreme sacrifice and altruism in which *Homo erectus* individuals died for the group, their genes passed on through their children and relatives who survived and reproduced. The nonverbalized altruism that motivates the most noble of *Homo sapiens* activities—from rescuing a drowning stranger to global philanthropy—probably derive from our *Homo erectus* heritage. But so do the unquestioning self-oblations of the kamikaze pilot or the suicide bomber. We have inherited from our Pleistocene ancestors powerful feelings, many of which we struggle to put into words because they evolved before our brains were wired for language. In a changed world of global interdependence, modern *Homo sapiens* has the cultural choice of extending and applying these ancient, beneficent emotions to the whole species, or of maintaining the primitive status quo of *Homo erectus*, an unreasoning, noninclusive, and xenophobic collective state of mind formerly termed "savagery" and "barbarism," or alternatively, "patriotism," "nationalism," or "tribalism."

Much of modern human behavior is learned, and that learning occurs within culture and during a relatively long childhood. *Homo erectus* is hypothesized to have been the first of our ancestors to show an increased duration of childhood, with accelerated growth rates during infancy and adolescence to compensate for the intervening period of slow down in physical growth. Analysis of the growth rates of tooth enamel in fossil hominids has been the primary evidence for this important deduction, and this conclusion needs to be tested and verified with other tests and more data. The sequence and rate of cultural change in *Homo erectus* are still poorly understood, and a goal of future research should be to relate these to physical growth changes in evolution.

In order to investigate cultural and biological adaptive changes in *Homo erectus* and how they subsequently evolved, hypotheses must be built on a firm substrate of population ecology. These parameters are the basic tenets of any species' adaptation—how it makes its living, where it spends its time on a daily basis, what it eats, what eats it, and how it reproduces. Dragon Bone Hill provided some of the first forays into these aspects of paleoanthropology.

Dragon Bone Hill Diet, Disease, and Ecology

The first archaeological sites in Europe, investigated as far back as the eighteenth century, were interpreted within the context of big game hunting. Teilhard de Chardin and Henri Breuil, the architects of archaeological interpretation at Zhoukoudian, were steeped within this tradition, and the first interpretations of Longgushan *Homo erectus* were along these lines. Indeed, it made sense to extend clear archaeological evidence of hunting of reindeer and other large mammals from geologically recent sites in Europe to earlier sites such as Longgushan in the absence of data to the contrary. As we have seen, in the 1930s Chinese researcher W. C. Pei first showed evidence to the contrary by disagreeing with the eminent Breuil over the presumed bone weapons and other tools found at Dragon Bone Hill. Many years later Lewis Binford's and our own research have confirmed that the bones that Breuil thought were hominid tools were in fact bone refuse from hyena meals. And although stone tools were found in abundance in the cave, none of them can be said to be sufficient or even adequate for killing large mammals on the hoof. Cut marks document that *Homo erectus* cut meat from large mammal carcasses but not that they hunted the animals down.

Our scenario for *Homo erectus* food getting supports those who have hypothesized that early hominids were scavengers,[16] but many questions still surround this fascinating and, to modern gustatory sensibilities, disgusting dietary adaptation. Primary among the questions that remain is whether fire, which we think was in the *Homo erectus* behavioral repertoire, was regularly used to cook meat or other foods, or whether it was used only as an ecological tool to displace other species and to burn away vegetation.

In their groundbreaking research on the use of fire at Longgushan, Steve Weiner and his colleagues determined conclusively that only already fossilized bone, which turned colors, had been burned in the fires that had ravaged the cave. Carbonized bone that Davidson Black had hypothesized to be evidence of charring were found to be coverings of finely laminated organic deposits laid down under water, probably in pools within the cave. The bone fragmentation thought by Lewis Binford to be due to burning or roasting horse heads remains the best evidence for any intentional cooking at Dragon Bone Hill. The exact relationships of *Homo erectus*, fire, and animal remains at the site are outstanding questions that need to be probed in much greater depth in the future.

General arguments about *Homo erectus*'s adaptation make a strong case for meat-eating in the species. The species' increase in body size correlates to an increased need for calories, which was required for covering more ground in its search for food, presumed because of theoretical expectations that larger animals occupy larger home ranges. Added to this, Leslie Aiello and her colleagues have made the convincing argument that as the brain enlarged in *Homo erectus*, the gut (stomach and intestines) decreased in size.[17] Known as the "expensive tissue hypothesis," this idea seems to be supported by anatomical changes in the *Homo erectus* skeleton that indicate that the abdominal area was smaller as were the chewing teeth and masticatory muscles. Because of the decreased food-handling and -processing anatomy, *Homo erectus* must have eaten substantially more high-quality food, that is, high-protein and high-calorie foods. Cut marks on bone that can only have been made by hominids confirm that meat, organs, and animal fat were a major component of this new, high-quality diet. But the range of meat, what species it came from, and when in the year it was eaten remain a mystery. Basic questions, such as how much meat was eaten, and by whom in the group, remain for future research. Was bone marrow eaten? How did Longgushan *Homo erectus* avoid getting too much vitamin A from eating carnivore liver? If these hominids ate too much meat, including purine-rich organs, did they ever get gout? Much needs to be explained about the meat component of the *Homo erectus* diet alone.

Longgushan was the first paleoanthropological site at which presumed evidence of *Homo erectus* plant foods was discovered. Ralph Chaney, a paleontologist and botanist from the University of California at Berkeley, investigated fossilized and burned seeds at Longgushan in the 1930s. He determined that the seeds known colloquially as hackberries, and widely eaten around the world, were from the *Celtis* tree. Chaney hypothesized that hackberries formed part of the diet of *Homo erectus*. As enticing as this idea may be, it is equally probable that the location of the hackberries in the site, which we have now determined was near the southern cave opening in the upper part of the deposit, is explicable by a *Celtis* tree's growing near the opening and dropping its berries into the cave. The burning of the berries, which Chaney had argued connected them with hominid activity, remains, but is attended with the same problem as the burned fossilized bone in the site. These problems await further research.

Two recent studies have emphasized the potential importance of tubers as food sources for *Homo erectus*. When we realize that there are no fossil remains of tubers or even indirect archaeological evidence that early hominids ever ate tubers (termed "USOs" or "underground storage organs" by specialists), we might question whether or not these ideas are to be taken seriously. There are some cogent arguments, however, that make the case for focused research even if they do not prove the case that tubers formed a major part of the diet of early *Homo*. Archaeologist James O'Connell of the University of Utah and his colleagues use data from modern African foragers (the Hadza of Tanzania) to argue that women's gathering of plant foods, particularly the //ekwa root,[18] is critical to survival in the often marginal habitats in which the Hadza live.[19] These authors relate this possible dietary adaptation specifically to *Homo erectus* and its spread out of Africa, terming it the "grandmother hypothesis." They suggest that post-menopausal women in the group could help feed children with the tubers they collected, and thus promote reproduction of their daughters and younger female relatives. The other study, headed by primatologist Richard Wrangham of Harvard University, starts from the observation that chimpanzee males are now known to hunt and then share meat, but these males do not provision nuclear families who live in "home bases," as the traditional-hunting hypothesis for *Homo erectus* maintained.[20] Chimpanzees do not share with *Homo erectus* many of the anatomical and physiological adaptations that allow the latter to eat and digest meat, so the analogy is somewhat strained. Chimps, for example, engage in coprophagy—picking partially digested meat out of their own feces to reingest it. Nevertheless, both models relate their hypotheses to the spread of open, savanna-like conditions, increased home-range sizes of *Homo erectus*, and the spread of hominids out of Africa. They both put a valid emphasis on the non-meat component of the *Homo erectus* diet, and, indeed, this must be better investigated.

Tests of dietary hypotheses are possible using isotopes of chemical elements, such as strontium, carbon-13, oxygen-18, and nitrogen-15 obtained from fossil teeth or bones. Michael Richards of Oxford University recently led one study that showed that Neandertals in Croatia regularly ate as much animal protein as large mammalian carnivores. This study analyzed a bone protein, collagen, which unfortunately is rarely, if ever, preserved in bone as old as *Homo erectus*. Other isotopes, however, that reside within the crystalline apatite structure of bones and teeth can be used to assess diet,[21] and should be employed to assess *Homo erectus* diet. Another promising area of dietary research is coprolite analysis. Coprolites are fossilized feces,[22] and there are hundreds, if not thousands, in the collections of the Institute of Vertebrate Paleontology and Paleoanthropology in Zhoukoudian and Beijing. The vast majority are presumed from their shape to be from hyenas, but analysis of even one *Homo erectus* coprolite that might be in the collection would provide a wealth of dietary information unknowable from any other source.

Diet and ecology naturally lead to questions about disease in *Homo erectus*. We have discussed a vitamin overload disease (hypervitaminosis A) and a parasitic disease (tapeworms) in *Homo erectus*. Archaeologists Ofer Bar-Yosef and Ann Belfer-Cohen have focused on *Homo erectus* disease in the context of the dispersion of earliest *Homo erectus* out of Africa and into Eurasia.[23] Citing Africa as a "garden of germs," these authors think that hominids could have escaped the parasite-infested and disease-laden tropics of Africa to expand into a much healthier environment in the colder north. They mention parasites, sleeping sickness, malaria, and elephantiasis as afflictions endemic to Africa that may have prompted population movements.

The late Pliocene time period, around two million years ago, was one in which several diseases may have become more prevalent in Africa, lending some credence to Bar-Yosef and Cohen's idea. At about this time, or slightly earlier, the lake snail *Bulinus* first appeared in the fossil record of eastern and central Africa.[24] This snail is one of the primary vectors of the human parasitic disease, schistosomiasis (also known as bilharzia), which may date from this time. Recent research indicates that malaria is a much more recent disease, with a protective genetic mutation (causing glucose-6-phosphate dehydrogenase deficiency) dating to only a few thousand years ago,[25] so at least this defense against the malarial parasite is not likely involved in this proposed mechanism. We have already seen that tapeworms seem to have diverged evolutionarily about 1.7 million years ago, but emigrating hominids did not escape them by going to Eurasia—the tapeworms rode along out of Africa within the hominids' and carnivores' guts. Sleeping sickness (trypansomiasis), caused by the bite of the tsetse fly, and elephantiasis, caused by the filarial worm (*Wucheria bancrofti*) are undated as to their probable earliest occurrence. Unmentioned by Bar-Yosef and Cohen are viruses, of which there are many deadly varieties in Africa. The presence of viruses affecting humans in Africa, especially in the forests, is probably ancient because of long coevolution with our ancestors and relatives there. HIV, and Ebola, Bunya, Semliki, West Nile, and Rift Valley Fever viruses are some of the more well known. Escape from endemic African viruses might well have been one advantage of the move out of Africa, particularly when population densities of early hominids were such that viral transmission from one population to another was limited. However, when *Homo erectus* left the African tropics, it soon spread to the Asian tropics. Here an abundant and diverse group of potentially infective parasitic, bacteriological, and viral diseases residing in primate populations from monkeys to apes exists, kept alive and well by warm, wet forests. One may well ask, if escape from disease was important to *Homo erectus*, why go from the African frying pan into the Asian fire? In addition, a whole new spate of diseases that tends to afflict European and Asian populations preferentially, such as cystic fibrosis, arose that would have counterbalanced the beneficial effects of leaving Africa. There are many questions still to be addressed in this interesting new arena of research, and molecular approaches may prove the most effective in investigating them.

Top: The Upper Cave at Longgushan, shown here, was discovered to the south of the main excavation of Locality 1. It was excavated between 1931 and 1933, and yielded a late Pleistocene fauna, abundant stone tools, and an example of modern *Homo sapiens*. *Bottom:* The most complete skull of *Homo sapiens* (no. 101) recovered from Upper Cave. This skull is fully modern in appearance. Unlike *Homo erectus*, Upper Cave *Homo sapiens* may have lived in the cave, documenting the remarkable persistence of hominids at Dragon Bone Hill for over half a million years.

Testing Theories on the Extinction
of *Homo erectus*

After being found in the uppermost strata of Dragon Bone Hill, dated at roughly 410,000 years ago, *Homo erectus* is no longer reliably recorded in mainland Asia. The next fossil hominids from China are skulls of *Homo heidelbergensis* from the sites of Jinniushan, approximately 280,000 years old; Dali, approximately 200,000 years old; and Mapa, approximately 132,000 years old. What transpired between the end of Dragon Bone Hill's time and the appearance of the earliest *Homo heidelbergensis* in China—a period of approximately 130,000 years—is a blank. *Homo heidelbergensis* in China, however, shares a number of *erectus*-like cranial features. Mainstream paleoanthropological opinion has held that these populations moved into China from western Eurasia or even Africa, replacing or living alongside *Homo erectus*. Actual temporal overlap between latest *Homo erectus* and earliest *Homo heidelbergensis* in China has never been demonstrated, however. Chinese paleoanthropological opinion tends to favor an in situ evolution of *Homo sapiens* from *Homo erectus*.[26] Recent dating of a modern *Homo sapiens* at the site of Liujiang in China—at least 68,000 years ago, and possibly older—has reasserted this position.[27] It is claimed that this specimen may be as old as anatomically modern *Homo sapiens* from Africa and the Levant, thereby lending support to an in situ evolution in China. Alternatively, Liujiang and its early date may be considered support for our clinal-replacement model, in which we would expect more or less synchronous appearances of new species worldwide.

The anatomically modern *Homo sapiens* discovered in the Upper Cave at Dragon Bone Hill is later in time than the Liujiang find, but it is anatomically similar and probably belonged to a closely related population. That population may have migrated to Dragon Bone Hill from Africa, as "Out of Africa" theorists maintain; it may have evolved in place, as Multiregionalist theorists maintain; or it may have replaced another slightly more primitive population as the last relay in a wave of replacement, as our clinal-replacement model suggests. The evolutionary continuity that seemed so obvious to Davidson Black and Franz Weidenreich between *Sinanthropus* and the Upper Cave *Homo sapiens* is far from agreed upon. Only more research will answer the question.

We moderns are wont to look at the past not only with a backward gaze, but down our noses. It is all too easy to regard *Homo erectus*, our parahuman, beetle-browed ancestor of long ago, as ultimately destined for extinction. After all, the species had subhuman intelligence and a cultural adaptation that changed more slowly than the glaciers that periodically descended on them from the north. Such a view misses the length of time that *Homo erectus* existed—1.5 million years, or 15 times as long as modern *Homo sapiens* has been around—and it ignores the vast advances the species made. The largest member of the genus *Homo* to have evolved up to its time, it also became the most widespread primate species in the world other than its own, later descendants. Despite the uniqueness of *Homo erectus* cranial anatomy—their skulls were bony carapaces buttressed by huge browridges and massive tori that protected their brains in aggressive encounters—and the inferences that they could not speak and did not have modern human levels of manual dexterity, the species nevertheless invented the stone biface, tamed fire, and learned to survive the fluctuations of the northern hemisphere's Ice Age. The unusual mix of capabilities that *Homo erectus* possessed signaled zoological success for the species. We inherited many of those anatomical and behavioral traits from *Homo erectus*, for good or ill, and we must bear in mind that the species became extinct by eventually evolving into ourselves.

Notes

Chapter 1   The Bones of Dragon Hill

1. "Hominid" refers to a member of the zoological family Hominidae, which in our usage includes humans and their bipedal predecessors. Some authors use the term to refer to apes and humans together, but we prefer the established term "hominoid" for this taxonomic grouping.
2. Schlosser, M. 1903. Die fossilen Säugethiere Chinas. *Abhandlungen der Bayerischen Akademie der Wissenschaften* II, 150: 22.
3. Osborn, H. F. 1924. American Men of the Dragon Bones. *Natural History*, 24(3): 350–65.
4. Ingersoll, E. 1928. *Dragons and Dragon Lore*. Payson & Clarke, New York.
5. Reader, J. 1981. *Missing Links, the Hunt for Earliest Man*. Little, Brown, Boston.
6. Andersson, J. G. 1925. Archaeological research in Kansu. *Chinese Geological Survey*, Series A, No. 5.
7. Andersson, J. G. 1928. *The Dragon and Foreign Devils*. Little, Brown, Boston.
8. Andersson, J. G. 1934. *Children of the Yellow Earth: Studies in Prehistoric China*. Kegan Paul, Trench, Trubner & Company, London.
9. Andersson (1934, p. 97).
10. Andersson (1934, p. 98).
11. Zdansky, O. 1928. Die Saeugetiere der Quataerfauna von Choukoutien. *Palaeontologia Sinica*, Series C, 5: 1–146.
12. Andersson (1934, p. 98).
13. Andersson (1934, p. 100).
14. Reader (1981, p. 104).
15. Reader (1981, p. 100).
16. Zdansky (1928).
17. Andersson (1934, p. 103).
18. Hood, D. 1964. *Davidson Black, A Biography*. University of Toronto Press, Toronto.
19. Hood (1964).

20. Andersson (1934, p. 189).
21. Black, D. 1928. A study of Kansu and Honan Aeneolithic skulls and specimens from Later Kansu prehistoric sites in comparison with North China and other recent crania. I. On measurement and identification. *Palaeontologia Sinica*, Series D, 6(1): 1–83.
22. Black, D. 1926. Tertiary man in Asia: The Choukoutien discovery. *Nature* 118: 733–34.
23. Andersson (1934, p. 1).
24. Andersson (1934, pp. 104–5).
25. Andersson (1934).
26. Hood (1964).
27. Black, D. 1927. The lower molar hominid tooth from the Chou Kou Tien deposit. *Palaeontologia Sinica*, Series D, 7(1): 1–28.
28. Dart, R. A. 1925. *Australopithecus africanus*, the man-ape of South Africa. *Nature* 115: 195–99.
29. Andersson (1934).
30. Black (1929).
31. Andersson (1934, p. 109).
32. Spencer, F. 1990. *Piltdown: A Scientific Forgery*. Oxford University Press, New York.
33. Pei, W. 1929. An account of the discovery of an adult *Sinanthropus* skull in the Choukoutien cave deposit. *Bulletin of the Geological Society of China* 8: 203–5.
34. Black, D. 1931a. On the adolescent skull of *Sinanthropus pekinensis* in comparison with an adult skull of the same species and with other hominid skulls, recent and fossil. *Palaeontologia Sinica*, Series D, 7(2): 1–144.
35. Hood (1964, p. 126).
36. Teilhard de Chardin, P. 1934. Letter to Walter Granger (March 19, 1934). Granger-Teilhard Collection, Georgetown University Library, letter 1: 11.
37. Teilhard de Chardin (1934).
38. Weidenreich was reportedly not religious and did not consider himself Jewish, but German "racial" policy held that anyone with one Jewish grandparent was classified as Jewish (Wolpoff, M., and R. Caspari. 1997. *Race and Human Evolution*. Simon & Schuster, New York.).
39. Gustav Schwalbe had studied under Johannes Müller, the great anatomist and embryologist, and was a contemporary of Rudolf Virchow. Schwalbe was a strong advocate of Darwinian human evolution and thus opposed Virchow's viewpoint. He was the leading contemporary expert on Neandertal anatomy and considered Neandertals ancestral to modern humans. Weidenreich was working in Schwalbe's lab when he was studying and describing Eugene Dubois's *Pithecanthropus* fossil skull from Java (Wolpoff and Caspari, 1997).
40. Gregory, W. K. 1949. Franz Weidenreich, 1873–1948. *American Anthropologist* 51: 85–90.
41. Gregory (1949).
42. Jia, L. (editor). 2000. *Chronicle of Zhoukoudian (1927–1937)*. Shanghai Scientific & Technical Publishers, Shanghai.
43. Teilhard de Chardin, P. 1935a. Letter to Walter Granger (March 27, 1935). Granger-Teilhard Collection, Georgetown University Library, letter 1: 17.
44. Teilhard de Chardin, P. 1935b. Letter to Walter Granger (July 25, 1935). Granger-Teilhard Collection, Georgetown University Library, letter 1: 13.

45. Jia (2000); see also Jia, L., and W. Huang. 1990. *The Story of Peking Man, from Archaeology to Mystery*. Oxford University Press, Oxford.
46. Teilhard de Chardin, P. 1936. Letter to Walter Granger (February 18, 1936). Granger-Teilhard Collection, Georgetown University Library.
47. Although the Cenozoic Research Laboratory at Peking Union Medical College ceased to exist administratively in 1941, the Institute of Vertebrate Paleontology and Paleoanthropology (IVPP) of the Chinese Academy of Sciences, which today undertakes and administers Chinese paleoanthropological research, was founded in 1949 to continue the work of the Cenozoic Research Laboratory. Dr. Zhongjian Yang, the longtime colleague and friend of Davidson Black and Teilhard de Chardin, became its first director.
48. In an internal report written by Dr. Warren Weaver, then director of the natural sciences of the Rockefeller Foundation (Weaver, W. 1941a. *Internal Report* (Friday, June 6, 1941), Rockefeller Foundation Archives, Sleepy Hollow, N.Y. Record Group 1.1, Series 601D, Box 39, Folder 323), he states: "there has been a major difference of opinion on policy between W[eidenreich] and Houghton. W[eidenreich] did not think it was at all necessary for him to leave China and obviously did so reluctantly and under definite orders from H[oughton]."
49. In a letter, dated April 10, 1941, to Mr. Edwin Lobenstine, chairman of the China Medical Board of the Rockefeller Foundation, Dr. Henry S. Houghton, president of Peking Union Medical College, writes: "Some weeks ago Dr. Weidenreich raised the question with me as to whether or not it might be possible or practicable, with the consent of the officials of the National Geological Survey and of the Chinese National Government, to remove the human material and artefacts to some one of the great museums in the United States, there to be held in custody for the duration of the war. After talking the matter over with him and with others interested, including the First Secretary of the Embassy, I came to the conclusion that it would not be in order to do so." (Houghton, H. S. 1941. Letter to Edwin Lobenstine [April 10, 1941]. Rockefeller Foundation Archives, Sleepy Hollow, N.Y. Record Group 1.1, Series 601D, Box 39, Folder 323).

Chapter 2 The Dragon Reclaims Its Own

1. Jia (2000).
2. Jia and Huang (1990, p. 152).
3. Jia and Huang (1990, p. 152).
4. Gibney, F. (editor) 1995. *Senso, The Japanese Remember the Pacific War*. M. E. Sharpe, Armonk, NY.
5. Jia and Huang (1990, p. 251).
6. Gibney (1995, p. 61).
7. Li, M. S., and N. Yue. 2000. *Search for Peking Man* [In Chinese]. Hua Xia Publishers, Beijing.
8. Weaver, W. 1941b. Notes on Interview by "WW" with Dr. F. Weidenreich (Friday, June 6, 1941). Rockefeller Foundation Archives, Sleepy Hollow, N.Y. Record Group 1.1, Series 601D, Box 39, Folder 323.
9. Ryuzo Torii (1870–1953) has been termed the "founding father of Japanese anthropology" and was active in archaeology, physical anthropology, and ethnography of

Asia. He was a contemporary of German-American anthropologist Franz Boas and shared with him a broad definition of the field.

10. Yenching University, the forerunner to Beijing University, was founded in 1916 as a missionary institution by the Presbyterian and Methodist churches. In 1952 it became Beijing University.

11. Weidenreich had been without official German citizenship since November 1935, when Nazi-controlled Germany passed the "Reich Citizenship Law" ["Reichsbürgergesetz"], which made German citizenship dependent upon German "blood," i.e., no descent from "non-Aryan" grandparents (see Proctor, R. 1988. From *Anthropologie* to *Rassenkunde* in the German anthropological tradition. In *Bones, Bodies, Behavior. Essays on Biological Anthropology*. History of Anthropology. Vol. 5. Edited by G. W. Stocking, pp. 138–79, University of Wisconsin Press, Madison, p. 160.). Weidenreich had applied for American citizenship but had "only his first papers" (noted in interview transcript by Warren Weaver, see note 8 above).

12. Jia and Huang (1990, p. 161).

13. Wong, W. H. (Weng, W. H.), and T. H. Yin. 1941. Letter to Dr. H. S. Houghton (January 10, 1941). Rockefeller Archives, Sleepy Hollow, N.Y. Record Group 1.1, Series 601D, Box 39, Folder 323.

14. Houghton, H. S. 1941. Letter to Edwin Lobenstine (April 10, 1941). Rockefeller Foundation Archives, Sleepy Hollow, N.Y. Record Group 1.1, Series 601D, Box 39, Folder 323.

15. Wong (Weng) and Yin (1941).

16. Houghton (1941).

17. Houghton (1941, p. 2).

18. See note 48 in chap. 1. In his letter of April 10, 1941, Houghton writes "Dr. Weidenreich, as you know, is proceeding at once to New York and will take with him this letter."

19. Weidenreich considered himself "stateless."

20. Weaver (1941b).

21. Weaver, W., and R. B. Fosdick. 1941. Notes on Interview by "WW" with "RBF" (Friday, June 6, 1941). Rockefeller Foundation Archives, Sleepy Hollow, N.Y. Record Group 1.1, Series 601D, Box 39, Folder 323.

22. Jia and Huang (1990, p. 160).

23. Li and Yue (2000).

24. Jia and Huang (1990, pp. 160–61).

25. Jia and Huang (1990).

26. Shapiro, H. 1974. *Peking Man: The Discovery, Disappearance and Mystery of a Priceless Scientific Treasure*. Simon & Schuster, New York.

27. Li and Yue (2000).

28. Beeman, C. L. 1959. Peking Man: Letters. *Science* 130: 416.

29. Lin, J. 1999. Mystery of the Missing Ancient Skulls. *Detroit Free Press*, May 17, 1999 (http://www.100megsfree4.com/farshores/skulls.htm).

30. Jia and Huang (1990, p. 161).

31. Boaz, N. T. 2002a. Personal communication from Martin Taschdjian, son of Claire Taschdjian, by telephone (December 12, 2002).

32. Li and Yue (2000, p. 367).

33. Janus, C., and W. Brashler. 1975. *The Search for Peking Man*. Macmillan, New York.

34. Boaz, N. T. 2002b. Personal communication from Christopher Janus by telephone (November 2002).
35. Interview with Mr. James Stewart-Gordon (Rasky, H. [editor] 1977. *The Peking Man Mystery*. Canadian Broadcasting Company, Toronto).
36. Hu, C. 1977. Letter to Lanpo Jia (March 4, 1977), reprinted in *The Story of Peking Man, from Archaeology to Mystery*. Edited by L. Jia and W. Huang. 1990. Oxford University Press, Oxford, pp. 160–61.
37. Interview with Wenzhong Pei, *Da Gong Bao* newspaper, Beijing, March 1950, reprinted in *The Story of Peking Man, from Archaeology to Mystery*. Edited by L. Jia and W. Huang. 1990. Oxford University Press, Oxford, pp. 171–72.
38. Jia and Huang (1990, p. 172).
39. Hasebe, K. 1915. Notes on the local customs of the Marshall Islands. *Journal of the Anthropological Society of Tokyo* 30(7): 278–79.
40. Fortuyn, A. B. D. 1942. Excerpt from report of Dr. Fortuyn on Department of Anatomy, P.U.M.C.—Part Two, dated December 5, 1941, with cover memorandum from "AMP" [Agnes M. Pearce, Secretary, China Medical Board, Inc.] (Rockefeller Archives, Sleepy Hollow, N.Y. Record Group 1.1, Series 601D, Box 39, Folder 323).
41. Jia and Huang (1990, p. 172).
42. Li and Yue (2000).
43. Jia and Huang (1990, p. 167).
44. Jia and Huang (1990, pp. 162–65).
45. Jia and Huang (1990, p. 165).
46. Interview with Dr. Lucian W. Pye (Rasky, 1977).
47. Taschdjian, C. 1977. *The Peking Man Is Missing*. Harper & Row, New York, p. 119.
48. Boaz (2002a).
49. Tobias, P., W. Qian, and J. L. Cormack. 2000. Davidson Black and Raymond Dart: Asian-African parallels in paleoanthropology. *Acta Anthropologica Sinica,* Supplement 19, Institute of Vertebrate Paleontology and Paleoanthropology, Beijing.

Chapter 3   Giants and Genes: Changing Views of Peking Man's Evolutionary Significance

1. Jia and Huang (1990).
2. Modern human cranial capacity is an average of about 1,350 cubic centimeters, ranging as high as 1,800 cc and as low as 1,000 cc. The measured cranial capacities of the Longgushan hominids range between 915 cc and 1,225 cc.
3. Weidenreich, F. 1939a. *Sinanthropus* and his significance for the problem of human evolution. *Bulletin of the Geological Society of China* 19: 1–17.
4. Weidenreich (1939a, p. 11).
5. Hrdlička, A. 1920. Shovel-shaped teeth. *American Journal of Physical Anthropology* 3: 429–65.
6. Smith, G. E. 1932. The discovery of primitive man in China. In *Smithsonian Report for 1931*. U.S. Government Printing Office, Washington, D.C., pp. 531–47.
7. Weidenreich, F. 1930. Ein neuer *Pithecanthropus* - Fund in China. *Natur und Museum* 60 (12): 546–51; see also chapter 1, note 39.

8. Ralph von Koenigswald was nominated for the position at the Dutch Geological Survey by Ferdinand Broili, one of his professors at Munich. Interestingly, Broili had been the mentor of Zhongjian Yang [C. C. Young], later one of the codirectors of the Longgushan excavations, who received his doctorate from Munich in 1927, one year before von Koenigswald. Although the two men must have been aware of, or even known, each other, there is no evidence of their having met or corresponded after they left Munich.

9. Koenigswald, G. H. R. von 1931a. *Sinanthropus, Pithecanthropus* en de ouderdom van de Trinil-Lagen. *Mijningenieur* 1931: 198–202.

10. Koenigswald, G. H. R. von 1931b. Fossilen uit Chineesche apotheen in West-Java. *Mijningenier* 1931: 189–93; Koenigswald, G. H. R. von 1932. Versteinerungen als Azneimittel bei den Chinesen auf Java. *Natur und Museum* 62 (9): 292–95.

11. Koenigswald, G. H. R. von 1936. Erste Mittelung über einen fossilen Hominiden aus dem Altpleistozän Ostjavas. *Natur und Museum* 62: 292–95; Koenigswald, G. H. R. von 1937. Ein Underkieferfragment des *Pithecanthropus* aus dem Trinil-schichten Mittlejavas. *Koninkl. Nederl. Akademie van Wetenschappen* 40: 883–93.

12. Koenigswald, G. H. R. von 1938. Ein neuer *Pithecanthropus*–Schädel. *Koninkl. Nederl. Akademie van Wetenschappen* 42: 185–92.

13. Weidenreich, F. 1940. Man or Ape? *Natural History* 45: 32–37.

14. Gustav Heinrich Ralph von Koenigswald was born in Berlin on November 13, 1902, and was not "Dutch" as sometimes averred (e.g., Walker, A., and P. Shipman. 1996. *The Wisdom of the Bones: In Search of Human Origins*. Alfred A. Knopf, New York). He went to gymnasium (secondary school) at Heppenheim an der Bergstrasse and studied at university at Berlin, Tübingen, Cologne, and Munich (where he received his doctorate in 1928). Between 1947 and 1968 von Koenigswald was indeed professor of paleontology at the Rijksuniversiteit Utrecht in the Netherlands, but then returned to Germany to the Senckenberg Museum in Frankfurt, where he remained until his death in 1982.

15. Boaz, N. T. 1975. Interview with G. H. R. von Koenigswald. Senckenberg Museum, Frankfurt.

16. Koenigswald, G. H. R. von, and F. Weidenreich. 1938. Discovery of an additional *Pithecanthropus* skull. *Nature* 142: 715.

17. Weidenreich, F. 1939b. The classification of fossil hominids and their relations to each other, with special reference to *Sinanthropus pekinensis*. *Bulletin of the Geological Society of China* 19: 64–75.

18. Koenigswald, G. H. R. von, and F. Weidenreich. 1939. The relationship between *Pithecanthropus* and *Sinanthropus*. *Nature* 144: 926–29.

19. Army officer Walter Fairservis had studied at Columbia University with physical anthropologist Dr. Harry Shapiro.

20. Weidenreich, F. 1943. The skull of *Sinanthropus pekinensis*: A comparative study on a primitive hominid skull. *Palaeontologia Sinica*, New Series D, 10: 1–485.

21. Weidenreich, F. 1941a. *The Brain and Its Role in the Phylogenetic Transformation of the Skull*. American Philosophical Society, Philadelphia.

22. Gregory (1949).

23. Ciochon, R., J. Olsen, and J. James. 1990. *Other Origins: The Search for the Giant Ape in Human Prehistory*. Bantam, Doubleday, New York.

24. "Multiregionalism" is a term coined by paleoanthropologist Milford Wolpoff of the University of Michigan (Wolpoff, M. 1999. *Paleoanthropology*. 2nd ed. McGraw–Hill, New York; see also Hawks, J., K. Hunley, S. Lee, and M. Wolpoff. 2000. Population bottlenecks and Pleistocene human evolution. *Molecular Biological Evolution* 17: 2–22).

25. Washburn, S. L., and D. Wolffson (editors). 1949. *The Shorter Anthropological Papers of Franz Weidenreich Published in the Period 1939–1948: A Memorial Volume*. The Viking Fund, New York.

26. Boaz, N. T. 1981. History of American paleoanthropological research on early Hominidae, 1925–1980. *American Journal of Physical Anthropology* 56: 397–406.

27. Rightmire, G. P. 1990. *The Evolution of* Homo erectus: *Comparative Anatomical Studies of an Extinct Human Species*. Cambridge University Press, Cambridge.

28. Isaac, G. L. 1975. Sorting out the muddle in the middle: An anthropologist's post-conference appraisal. In *After the Australopithecine*. Edited by K. Butzer and G. L. Isaac, Mouton, The Hague, pp. 875–87.

29. Dobzhansky, T. 1937. *Genetics and the Origin of Species*. Columbia University Press, New York.

30. Washburn, S. L. 1983. Evolution of a teacher. *Annual Review of Anthropology* 12: 1–24. Reprinted in *The New Physical Anthropology: Science, Humanism, and Critical Reflection*. Edited by Shirley Strum, Donald G. Lindburg, and David Hamburg. Prentice Hall, Upper Saddle River, New Jersey, pp. 215–27.

31. Washburn, S. L. 1946. The effect of facial paralysis on the growth of the skull of rat and rabbit. *Anatomical Record* 94: 163–68; Washburn, S. L. 1947. The relation of the temporal bone to the form of the skull. *Anatomical Record* 99: 239–48.

32. Weidenreich (1941a).

33. Weidenreich, F. 1946. *Apes, Giants, and Man*. University of Chicago Press, Chicago, p. 89.

34. Dobzhansky, T. 1942. Races and methods of their study. *Transactions of the New York Academy of Sciences* 4: 115–33.

35. Weidenreich (1946, p. 3).

36. Simpson, G. G. 1945. The principles of classification and the classification of mammals. *Bulletin of the American Museum of Natural History* 85: 1–350.

37. Washburn, S. 1951. The new physical anthropology. *Transactions of the New York Academy of Sciences* 13: 298–304.

38. Washburn, S. 1964. The origin of races: Weidenreich's opinion. *American Anthropologist* 66: 1165–67.

Chapter 4   The Third Function: A Hypothesis on the Mysterious Skull of Peking Man

1. U.S. Department of Health and Human Services. 2002. Deaths: Leading Causes for 2000. *National Vital Statistics Reports* 50(16): 1–86.

2. It is of interest that Weidenreich published several papers on African australopithecines and their significance to human evolution early in his career. His overall opinion, however, especially after he took over the interpretive role for Longgushan

Homo erectus, was that australopithecines were clearly on the ape side of the human–ape evolutionary divide. In his 1943 monograph, there is a striking absence of any mention of *Australopithecus*. Weidenreich only briefly discusses the African late Pleistocene *Homo sapiens* Olduvai Hominid 1 and "*Africanthropus*" fossils from Laetoli.

3. Brown, P. 1994. Cranial vault thickness in Asian *Homo erectus* and *Homo sapiens*. *Courier Forschungs-Institut Senckenberg* 17: 33–46.

4. Taplin, A. 1874. *The Narrinyeri: An Account of the Tribes of South Australian Aborigines Inhabiting the Country Around the Lakes Alexandrina, Albert, and Coorong, and the Lower Part of the River Murray*. E. S. Wigg & Son, Adelaide.

5. Brown (1994, p. 42).

6. An early record of thickened cranial bones due to malaria in *Homo sapiens* was discovered at the Ishango archaeological site, Democratic Republic of Congo (Boaz, N. T., P. Pavlakis, and A. S. Brooks. 1990. Late Pleistocene–Holocene human remains from the Upper Semliki, Zaire. In *Evolution of Environments and Hominidae in the African Western Rift Valley*. Edited by N. T. Boaz. Virginia Museum of Natural History, Memoir No. 1, pp. 3–14).

7. Bartsiokas, A. 2002. Hominid cranial bone structure: A histological study of Omo 1 specimens from Ethiopia using different microscopic techniques. *Anatomical Record* 267: 52–59.

8. LeCount, E. R., and C. W. Apfelbach. 1920. Pathologic anatomy of traumatic fractures of the cranial bones and concomitant brain injuries. *Journal of the American Medical Association* 74: 501–11.

9. Boaz, N. T. 1999. Personal observation. NTB, Bosnia Project, Physicians for Human Rights.

10. Le Fort, R. 1901. Étude experimentale sur les fractures de la machoire supérieure. *Revue de Chirurgie*. 23: 201–10.

11. Weidenreich, F. 1951. Morphology of Solo Man. *Anthropological Papers of the American Museum of Natural History* 43: 205–90.

12. Hawks, J. 2003. The browridge: Pleistocene body armor. *American Journal of Physical Anthropology*, Supplement 36: 112.

13. Weidenreich, F. 1938. The ramification of the middle meningeal artery in fossil hominids and its bearing upon phylogenetic problems. *Palaeontologia Sinica*, New Series D, 3: 1–16.

14. Weidenreich (1943).

15. Rogers, L. F. 1992. *Radiology of Skeletal Trauma*. Churchill Livingstone, New York.

16. Falk, D. 1992. *Braindance*. Henry Holt & Company, New York.

Chapter 5   The Adaptive Behavior of the Not-Quite-Human

1. Black, D. 1930. On an adolescent skull of *Sinanthropus pekinensis* in comparison with an adult skull of the same species and with other hominid skulls, recent and fossil. *Palaeontologia Sinica* Series D, 7, p. 208 [quoted by Pei, W., and S. Zhang. 1985. *A Study of the Lithic Artifacts of* Sinanthropus. Science Press, Beijing, p. 263].

2. Pei, W., and S. Zhang. 1985. *A Study of the Lithic Artifacts of* Sinanthropus. Beijing: Science Press, p. 263.

3. Jia (2000, p. 137).

4. Pei, W. 1931a. Notice of the discovery of quartz and other stone artifacts in the Lower Pleistocene hominid-bearing sediments of the Choukoutien cave deposit. *Bulletin of the Geological Society of China* 11: 109–46.

5. Teilhard de Chardin, P., and W. Pei. 1932. The lithic industry of the *Sinanthropus* deposits in Choukoutien. *Bulletin of the Geological Society of China* 11: 317–58.

6. Teilhard de Chardin, P. 1941. *Early Man in China*, No. 7. Institut de Géo-Biologie, Pékin.

7. Teilhard de Chardin (1941, p. 60).

8. Breuil, H. 1939. Bone and Antler Industry of the Choukoutien *Sinanthropus* Site. *Palaeontologia Sinica* 117: 1–93.

9. Breuil's idea of bone, tooth, and antler tools as precursors to stone tools formed the basis of Raymond Dart's later but much more widely known "osteodontokeratic" culture, which he proposed for the australopithecines at Makapansgat, South Africa, in the late 1930s. Taphonomic studies carried out primarily by C. K. Brain (Brain, C. 1981. *The Hunters or the Hunted? An Introduction to African Cave Taphonomy.* University of Chicago Press, Chicago) showed that these presumed bone tools were in fact remains of carnivore, especially hyaenid, modification.

10. Breuil (1939, p. i).

11. Breuil (1939, p. 4).

12. Breuil (1939, Plate 19, pp. 78–79).

13. Binford, L. R., and C. K. Ho. 1985. Taphonomy at a distance: Zhoukoudian, "the cave home of Beijing man?" *Current Anthropology* 26: 413–42; Binford, L. R., and N. M. Stone. 1986. Zhoukoudian: A closer look. *Current Anthropology* 27: 453–75.

14. Black, D. 1931b. Evidences of the use of fire by *Sinanthropus. Bulletin of the Geological Society of China* 11: 107–98.

15. Breuil, H. 1931. Le feu et l'industrie lithique et osseuse à Choukoutien. *Bulletin of the Geological Society of China* 11: 147–54.

16. Goldberg, P., S. Weiner, O. Bar-Yosef, Q. Wu, and J. Liu. 2001. Site-formation processes at Zhoukoudian, China. *Journal of Human Evolution* 41: 483–530.

17. Weiner, S., Q. Q. Xu, P. Goldberg, J. Y. Liu, and O. Bar-Yosef. 1998. Evidence for the use of fire at Zhoukoudian, China. *Science* 281: 251–53.

18. Pope, G. G. 1993. Ancient Asia's cutting edge. *Natural History* 5: 55–59.

19. Movius, H. L. 1969. Lower Paleolithic archaeology in southern Asia and the Far East. In *Studies in Physical Anthropology: Early Man in the Far East.* Edited by W. W. Howells, pp. 17–77. Anthropological Publications, The Netherlands.

20. Schick, K. D., N. S. Toth, W. Qi, J. D. Clark, and D. Etler. 1991. Archaeological perspectives in the Nihewan Basin, China. *Journal of Human Evolution* 27: 13–26.

21. Hopwood, D. E. 2003a. Behavioural differences in the early to mid-Pleistocene: Were African and Chinese *Homo erectus* really that different? *American Journal of Physical Anthropology*, Supp 36: 117–18.

22. Clark, J. D., and J. W. K. Harris. 1986. Fire and its roles in early hominid lifeways. *African Archaeological Review* 3: 3–29.

23. Rowlett, R. 1999. Did the use of fire for cooking lead to a diet change that resulted in the expansion of brain size in *Homo erectus* from that of *Australopithecus africanus? Science* 283: 2005.

24. See Howell, F. C. 1965. *Early Man.* Time-Life, New York.

25. Hoberg, E., N. L. Alkire, A. de Queiroz, and A. Jones. 2001. Out of Africa: Origins of the *Taenia* tapeworms in humans. *Proceedings of the Royal Society of Biological Sciences* 268(1469): 781–87.
26. Milius, S. 2001. Tapeworms tell tales of deeper human past. *Science News* 159: 215–16.

Chapter 6   The Times and Climes of *Homo erectus*

1. Teilhard de Chardin (1941).
2. Zhao, S., et al. 1985. Uranium-series dating of Peking Man site. In *Multi-Disciplinary Study of the Peking Man Site at Zhoukoudian*. Edited by R. Wu et al. Science Press, Beijing, pp. 246–55.
3. Yuan, S. X., T. M. Chen, S. J. Gao, and Y. Q. Hu. 1991. Study on uranium series dating of fossil bones from Zhoukoudian. *Acta Anthropologica Sinica* 10: 189–93.
4. Shen, G., and L. Jin. 1993. Restudy of the upper age limit of Beijing man site. *International Journal of Anthropology* 8: 95–98.
5. Qian, F., J. Zhang, and J. Li. 1980. Magnetostratigraphic study of the cave deposit containing fossil Peking Man at Zhoukoudian [In Chinese]. *Kexue Tong Bao* 25: 359.
6. Zhou, C., Z. Liu, Y. Wang, and Q. Huang. 2000. Climatic cycles investigated by sediment analysis in Peking Man's cave, Zhoukoudian, China. *Journal of Archaeological Science* 27: 101–9.
7. Goldberg et al. (2001).
8. Zhou et al. (2000).
9. Xie, Y., et al. 1985. The sedimentary environment of the Peking Man period. In *Multi-Disciplinary Study of the Peking Man Site at Zhoukoudian*. Edited by R. Wu et al. Science Press, Beijing, pp. 185–215.
10. Turner, C. 2003. When people fled hyenas: Oversized hyenas may have delayed human arrival in North America (review by Lee Dye) http://abcnews.go.com/sections/scitech/DyeHard/dyehard021120.html (accessed June, 2003).
11. Spencer, L. M. 1997. Dietary adaptations of Plio-Pleistocene Bovidae: Implications for hominid habitat use. *Journal of Human Evolution* 32: 201–28.
12. Boaz, N. T. 1979. Early hominid population densities: New estimates. *Science* 206: 592–95; Aiello, L. C., and J. Wells. 2002. Energetics and the evolution of the genus *Homo*. *Annual Review of Anthropology* 31: 323–38.
13. Walker, A., and R. E. Leakey. 1993. *The Nariokotome* Homo erectus *Skeleton*. Harvard University Press, Cambridge.
14. Isbell, L. A., J. D. Pruetz, M. Lewis, and T. P. Young. 1998. Locomotor activity differences between sympatric patas monkeys (*Erythrocebus patas*) and vervet monkeys (*Cercopithecus aethiops*): Implications for the evolution of long hindlimb length in *Homo*. *American Journal of Physical Anthropology* 105: 199–207.
15. Chaney, R. 1935. The food of "Peking Man." *Carnegie Inst. Washington, News Serv. B*, 3 (25): 198–202.

Chapter 7   The Nature of Humanness at Longgushan:
Brain, Language, Fire, and Cannibalism

1. Microcephaly, defined statistically and clinically as a cranial capacity two standard deviations below the population mean (Opitz, J. M., and M. C. Holt. 1990. Microcephaly: General considerations and aids to nosology. *Journal of Craniofacial Genetics and Developmental Biology* 10(2): 175–204), may have diverse causes and is associated with many clinically recognized syndromes. It is frequently associated with poor development of the brain, congenital malformations, and mental retardation, and may be inherited on the X-chromosome or autosomally, as either a dominant or recessive trait (see Online Mendelian Inheritance in Man. http://www.ncbi.nlm.nih.gov.)

2. Keith, A. 1927. The brain of Anatole France. *British Medical Journal* 2: 1048–49.

3. Black, D., P. Teilhard de Chardin, C. C. Young, and W. C. Pei. 1933. Fossil man in China: The Choukoutien cave deposits with a synopsis of our present knowledge of the late Cenozoic in China. *Memoirs of the Geological Survey of China*, Beijing, Series A, No. 11.

4. Weidenreich, F. 1936a. Observations on the form and the proportions of the endocranial casts of *Sinanthropus pekinensis*, other hominids, and the great apes: A comparative study of brain size. *Palaeontologia Sinica*, New Series D, 7 (4): 1–50.

5. Laitman, J., and R. C. Heimbuch. 1982. The basicranium of Plio-Pleistocene hominids as an indicator of their upper respiratory systems. *American Journal of Physical Anthropology* 59: 323–43.

6. Kay, R., M. Cartmill, and M. Balow. 1998. The hypoglossal canal and the origin of human vocal behavior. *Proceedings of the National Academy of Sciences, USA* 95: 5417–19.

7. DeGusta, D., W. H. Gilbert, and S. P. Turner. 1998. Hypoglossal canal size and hominid speech. *Proceedings of the National Academy of Sciences, USA* 96: 1800–4.

8. MacLarnon, A. 1993. The vertebral canal. In *The Nariokotome Homo erectus Skeleton*. Edited by A. Walker and R. Leakey. Harvard University Press, Cambridge, pp. 359–90.

9. Krantz, G. S. 1980. Sapienization and speech. *Current Anthropology* 21: 773–92.

10. Klein, R. G., and B. Edgar. 2002. *The Dawn of Human Culture*. John Wiley & Sons, New York.

11. Boesch, C., and H. Boesch. 1990. Tool use and tool making in wild chimpanzees. *Folia Primatologica* 54: 86–99.

12. Breuil, H. 1939. Bone and antler industry of the Choukoutien *Sinanthropus* site. *Palaeontologia Sinica* New Series D, 6: 1–93.

13. Boaz, N. T., R. L. Ciochon, Q. Xu, and J. Liu. 2000. Large mammalian carnivores as a taphonomic factor in the bone accumulation at Zhoukoudian. *Acta Anthropologica Sinica*, Supplement 19: 224–34.

14. Villa, P. 1992. Cannibalism in Prehistoric Europe. *Evolutionary Anthropology* 3: 93–104; Villa, P., C. Bouville, J. Courtin, D. Helmer, E. Mahieu, P. Shipman, G. Belluomini, and M. Branca. 1986. Cannibalism in the Neolithic. *Science* 233: 431–37; White, T. 1992. *Prehistoric Cannibalism at Mancos 5MTUMR-2346*. Princeton University Press, Princeton.

15. Asfaw, B., T. D. White, O. Lovejoy, B. Latimer, S. Simpson, and G. Suwa. 1999. *Australopithecus garhi*: A new species of early hominid from Ethiopia. *Science* 284: 629–35.

16. Fernandez, Y., J. C. Diez, I. Caceres, and J. Rosell. 1999. Human cannibalism in the early Pleistocene of Europe (Gran Dolina, Sierra de Atapuerco, Burgos, Spain). *Journal of Human Evolution* 37: 591–622.

17. "Wir können daher wohl mit Recht als eine besondere (21ste) Stufe unserer menschlichen Ahnenreihe den sprachlosen Menschen (Alalus) oder Affenmenschen (*Pithecanthropus*) unterscheiden, welcher zwar körperlich dem Menschen in allen wesentlichen Merkmalen schon gleichgebildet, aber noch ohne den Besitz der gegliederten Wortsprache war." (Haeckel, E. 1868. *Natürliche Schöpfungsgeschichte.* Fischer, Jena).

18. Walker, A., M. R. Zimmerman, and R. E. Leakey. 1982. A possible case of hypervitaminosis A in *Homo erectus. Nature* 296: 248–50.

19. A more recent interpretation of the pathological bone deposits in the *Homo erectus* skeleton ER 1808 is that this individual had yaws, a parasitic infection by spirochete bacteria (Rothschild, B. M., I. Hershkovitz, and C. Rothschild. 1995. Origin of yaws in the Pleistocene. *Nature* 378: 343). If so, this would be by far the earliest evidence of this disease in the world. Yaws tends to afflict modern children between the ages of two and five years old and it is most common in overcrowded conditions in the hot and humid tropics. Although possible, this interpretation is less likely for an adult hominid in the sparsely populated mid-Pleistocene in an area of northern Kenya not likely to have been "humid" at this time.

20. O'Connell, J., K. Hawkes, and N. G. Blurton Jones. 1999. Grandmothering and the evolution of *Homo erectus. Journal of Human Evolution* 36: 461–86.

21. Elgar, M. A., and B. J. Crespi (editors). 1992. *Cannibalism: Ecology and Evolution Among Diverse Taxa.* Oxford University Press, Oxford.

22. Wynn, T. 1993. Two developments in the mind of *Homo erectus. Journal of Anthropological Archaeology* 12: 299–322.

23. Wynn (1993, p. 299).

Chapter 8   Alpha and Omega: Resolving the Ultimate Questions of the Beginnings and Endings of *Homo erectus,* the Species

1. We use the term "hominid" in its generally accepted sense of a member of the family Hominidae, composed of humans and their bipedal fossil relatives. Some specialists use the term "hominin" to refer to this grouping, which is a taxonomic term for a zoological tribe. In our usage, hominids are a separate family distinct from the families of apes—the ape and human families are included in the superfamily Hominoidea.

2. For a summary of hominoid evolution, see Boaz, N. T., and A. J. Almquist. 2002. *Biological Anthropology,* 2nd ed. Prentice Hall, Upper Saddle River, N.J.; and Fleagle, J. G. 1998. *Primate Evolution and Adaptation,* 2nd ed. Academic Press, London.

3. See Boaz, N. T. 1997a. *Eco Homo.* Basic Books, New York.

4. Tobias, P., and G. H. R. von Koenigswald. 1964. A comparison between the Olduvai hominines and those of Java and some implications for hominid phylogeny. *Nature* 204: 515–18.

5. Boaz, N. T., and F. C. Howell. 1977. A gracile hominid cranium from upper Member G of the Shungura Formation, Ethiopia. *American Journal of Physical Anthropology* 46: 93–108.

6. Cronin, J., N. T. Boaz, C. B. Stringer, and Y. Rak. 1981. Tempo and mode in hominid evolution. *Nature* 292: 113–22.

7. Tobias, P. V. 1991. *Olduvai Gorge, Vol. 4. The Skulls, Endocasts, and Teeth of* Homo habilis. Cambridge University Press, Cambridge.

8. Brown, F. 1994. Development of Pliocene and Pleistocene chronology of the Turkana Basin, East Africa, and its relation to other sites. In *Integrative Paths to the Past: Paleoanthropological Advances in Honor of F. Clark Howell*. Edited by R. S. Corruccini and R. L. Ciochon. Prentice Hall, Englewood Cliffs, New Jersey, pp. 285–312.

9. Sartono, S. 1971. Observations on a new skull of *Pithecanthropus* erectus (*Pithecanthropus* VIII) from Sangiran, Central Java. *Courier Forschungs-Institut Senckenberg* 74(2): 185–94.

10. This specimen was named *Homo modjokertensis* by von Koenigswald in 1936 and was recently considered early *Homo erectus* in a study by Susan Anton (Anton, S. 2002. Evolutionary significance of cranial variation in Asian *Homo erectus*. *American Journal of Physical Anthropology* 118: 301–23).

11. Swisher III, C. C., G. H. Curtis, T. Jacob, A. G. Getty, A. Suprijo, and Widiasmoro. 1994. Age of the earliest known hominids in Java, Indonesia. *Science* 266: 1118–21.

12. Huang, W., R. L. Ciochon, Y. Gu, R. Larick, Q. Fang, C. Yonge, J. de Vos, H. Schwarcz, and W. Rink. 1995. Early *Homo* and associated artefacts from Asia. *Nature* 378: 275–78.

13. Gabunia, L., A. Vekua, and D. Lordkipanidze. 2000. The environmental contexts of early human occupation of Georgia (Transcaucasia). *Journal of Human Evolution* 38(2): 785–802.

14. Groves, C. P., and V. Mazak. 1975. An approach to the taxonomy of the Hominidae: Gracile Villafranchian hominids of Africa. *Casopis pro Mineralogii a Geologii* 20: 225–47.

15. Boaz (1997a).

16. Larick, R., and R. L. Ciochon. 1996a. The African emergence and early Asian dispersals of the genus *Homo*. *American Scientist* 84: 538–51.

17. Boaz, N. T. 1997b. Calibration and extension of the record of Plio-Pleistocene Hominidae. In *Biological Anthropology: The State of the Science*. Edited by N. T. Boaz and L. Wolfe. 2nd ed. International Institute of Human Evolutionary Research, Bend, Oregon, pp. 25–52.

18. Cann, R. L., M. Stoneking, and A. C. Wilson. 1987. Mitochondrial DNA and human origins. *Nature* 325: 31–36.

19. Eller, E. 2002. Population extinction and recolonization in human demographic history. *Mathematical Biosciences* 177/178: 1–10.

20. Stringer, C. B., and R. McKie. 1997. *African Exodus: The Origins of Modern Humanity*. Henry Holt, New York.

21. Ruff, C. 1991. Climate and body shape in hominid evolution. *Journal of Human Evolution* 21: 81–105.

22. Zuckerkandl, E., and L. Pauling. 1965. Molecules as documents of evolutionary history. *Journal of Theoretical Biology* 8(2): 357–66.

23. See Strum, S. C., D. Lindburg, and D. Hamburg (editors). 1999. *The New Physical Anthropology: Science, Humanism, and Critical Reflection.* Prentice Hall, Upper Saddle River, N.J.

24. Aiello and Wells (2002); Anton, S. C., W. R. Leonard, and M. L. Robertson. 2002. An ecomorphological model of the initial hominid dispersal from Africa. *Journal of Human Evolution* 43: 773–85.

25. Huffman, O. F. 2001. Geologic context and age of the Perning/Mojokerto *Homo erectus*, East Java. *Journal of Human Evolution* 40: 353–62.

26. Anton (2002).

27. The nickname of the "Out of Africa" theory comes from the novel of the same name written by Danish writer Isak Dinesen (Dinesen, I. [K. Blixen]. 1937. *Out of Africa.* Putnam, London). It was made into a motion picture in 1985, was directed by Sydney Pollack, and starred Meryl Streep and Robert Redford.

28. Swisher III, C. C., W. J. Rink, S. C. Anton, H. P. Schwarcz, G. H. Curtis, A. Suprijo, and Widiasmoro. 1996. Latest *Homo erectus* in Java: Potential contemporaneity with *Homo sapiens* in Southeast Asia. *Science* 274: 1870–74.

29. Rightmire, G. P. 1998. Human evolution in the Middle Pleistocene: The role of *Homo heidelbergensis. Evolutionary Anthropology* 6: 218–27.

30. Boaz (1997a).

31. Heslop, D., M. J. Dekkers, and C. G. Langereis. 2002. Timing and structure of the mid-Pleistocene transition: Records from the loess deposits of northern China. *Palaeogeography, Palaeoclimatology, Palaeoecology.* 185: 133–43.

32. Swisher et al. (1996).

33. Templeton, A. 2002. Out of Africa again and again. *Nature* 416: 45–51.

Chapter 9   Testing the New Hypotheses

1. Zhu, R., Y. X. Pan, B. Guo, C. D. Shi, Z. T. Guo, B. Y. Yuan, Y. M. Hou, W. W. Huang, K. A. Hoffman, R. Potts, and C. L. Deng. 2001. Earliest presence of humans in northeast Asia. *Nature* 413: 413–17.

2. Cronin, et al. (1981).

3. Krings, M., H. Geisert, R. W. Schmitz, H. Krainitzki, and S. Paabo. 1999. DNA sequence of the mitochondrial hypervariable region II from the Neandertal-type specimen. *Proceedings of the National Academy of Sciences* 96: 5581–85.

4. Simpson, G. G. 1964. *Tempo and Mode in Evolution.* Hafner, New York.

5. Boaz, N. T., D. Ninkovich, and M. Rossignol-Strick. 1982. Paleoclimatic setting for *Homo sapiens neanderthalensis. Naturwissenschaften* 69: 29–33.

6. Arsuaga, J. L., I. Martinez, C. Lorenzo, A. Gracia, A. Muñoz, O. Alonso, and J. Gallego. 1999. The human cranial remains from Gran Dolina Lower Pleistocene site (Sierra de Atapuerca, Spain). *Journal of Human Evolution* 37: 431–57.

7. McCown, T. D., and A. Keith. 1937–39. *The Stone Age of Mount Carmel. II. The Fossil Human Remains from the Levalloiso-Mousterian.* Clarendon, Oxford.

8. Bar-Yosef, O. 1998. Early colonizations and cultural continuities in the Lower Paleolithic of Western Asia. In *Early Human Behavior in Global Context: The Rise and*

Diversity of the Lower Palaeolithic Record. Edited by M. Petraglia and R. Korisettar. Routledge, London, pp. 221–79.

9. Flint, R. F. 1971. *Glacial and Quaternary Geology.* Wiley, New York.

10. Smith, F., E. Trinkaus, P. B. Pettitt, I. Karavanic, and M. Paunovic. 1999. Direct radiocarbon dates for Vindija G_1 and Velika Pecina Late Pleistocene hominid remains. *Proceedings of the National Academy of Sciences* 96(22): 12281–86.

11. Duarte, C., J. Mauricio, P. B. Pettitt, P. Souto, E. Trinkaus, H. van der Plicht, and J. Zilhao. 1999. The early Upper Paleolithic human skeleton from Abrigo do Lagar Velho (Portugal) and modern human emergence in Iberia. *Proceedings of the National Academy of Sciences* 96: 7604–9.

12. Stringer, C. B., and R. McKie. 1997.

13. Bar-Yosef, O., and A. Belfer-Cohen. 2001. From Africa to Eurasia—Early dispersals. *Quaternary International* 75: 19–28.

14. Brown (1994).

15. Ardrey, R. 1961. *African Genesis; A Personal Investigation into the Nature of Man.* Atheneum, New York.

16. Blumenshine, R., J. A. Cavallo, and S. D. Capaldo. 1994. Competition for carcasses and early hominid behavioral ecology: A case study and conceptual framework. *Journal of Human Evolution* 27: 197–213.

17. Aiello, L. C., and P. L. Wheeler. 1995. The expensive tissue hypothesis: The brain and the digestive system in human and primate evolution. *Current Anthropology* 36: 199–221.

18. The phoneme "//" represents a "click" (pronounced with an intake of air) in the Hadza language.

19. O'Connell et al. (1999).

20. Wrangham, R., J. H. Jones, G. Laden, D. Pilbeam, and N. Conklin-Brittain. 1999. The raw and the stolen: Cooling and the ecology of human origins. *Current Anthropology* 5: 567–94.

21. Sillen, A., G. Hall, and R. Armstrong. 1998. $^{87}Sr/^{86}Sr$ ratios in modern and fossil food-webs of the Sterkfontein Valley: Implications for early hominid habitat preferences. *Geochimica et Cosmochimica Acta* 62: 2463–78.

22. Jouy-Avantin, F., A. Debenath, A.-M. Moigne, and H. Moné. 2003. A standardized method for the description and the study of coprolites. *Journal of Archaeological Science* 30(3): 367–72.

23. Bar-Yosef, O., and A. Belfer-Cohen. 2000. Early human dispersals: The unexplored constraint of African diseases. In *Early Humans at the Gates of Europe.* Edited by D. Lordkipanidze, O. Bar-Yosef, and M. Otte. Proceedings of the First International Symposium, Dmanisi, Tbilisi, Georgia. *Etudes et Recherches Archaeologiques de l'Université de Liège* 92: 79–86.

24. Williamson, P. G. 1981. Paleontological documentation of speciation in Cenozoic mollusks from Turkana Basin. *Nature* 293: 437–43.

25. Tishkoff, S. A., R. Varkonyi, N. Cahinhinan, S. Abbes, G. Argyropoulos, G. Destro-Bisol, A. Drousiotou, B. Dangerfield, G. Lefranc, J. Loiselet, A. Piro, M. Stoneking, A. Tagarelli, G. Tagarelli, E. Touma, S. Williams, and A. Clark. 2001. Haplotype diversity and linkage disequilibrium at human *G6PD*: Recent origin of alleles that confer malarial resistance. *Science* 293: 455–62.

26. Wu, X. 1989. Early *Homo sapiens* in China. In *Early Humankind in China*. Edited by X. Wu and S. Shang. Science Press, Beijing.
27. Shen, G., W. Wang, Q. Wang, J. Zhao, K. Collerson, C. Zhou, and P. V. Tobias. 2002. U-series dating of Liujiang hominid site in Guangxi, southern China. *Journal of Human Evolution* 43(6): 817–29.

Bibliography

Aiello, L. C., and J. Wells. 2002. Energetics and the evolution of the genus *Homo*. *Annual Review of Anthropology* 31: 323–38.

Aiello, L. C., and P. L. Wheeler. 1995. The expensive tissue hypothesis: The brain and the digestive system in human and primate evolution. *Current Anthropology* 36: 199–221.

Aigner, J. 1987. Correlations ¹⁸O et Localité 1 de Chou-Kou-Tien. *L'Anthropologie* 91: 733–48.

Ambrose, S. 2001. Paleolithic technology and human evolution. *Science* 291: 1748–53.

Andersson, J. G. 1925. Archaeological research in Kansu. *Chinese Geological Survey*, Series A, No. 5.

———. 1928. *The Dragon and Foreign Devils*. Little, Brown, Boston.

———. 1934. *Children of the Yellow Earth: Studies in Prehistoric China*. Kegan Paul, Trench, Trubner & Company, London.

———. 1943. Researches into the prehistory of the Chinese. *The Museum of Far Eastern Antiquities (Östasiatiska Samlingarna) Stockholm*, Bulletin 15: 1–304.

Anton, S. 2002. Evolutionary significance of cranial variation in Asian *Homo erectus*. *American Journal of Physical Anthropology* 118: 301–23.

Anton, S., and K. J. Weinstein. 1999. Artificial cranial deformation and fossil Australians revisited. *Journal of Human Evolution* 36: 195–209.

Anton, S. C., W. R. Leonard, and M. L. Robertson. 2002. An ecomorphological model of the initial hominid dispersal from Africa. *Journal of Human Evolution* 43: 773–85.

Anton, S., S. Marquez, and K. Mowbray. 2002. Sambungmacan 3 and cranial variation in Asian *Homo erectus*. *Journal of Human Evolution* 43: 555–62.

Ardrey, R. 1961. *African Genesis; A Personal Investigation into the Nature of Man*. Atheneum, New York.

Arribas, A., and P. Palmqvist. 1998. Taphonomy and paleoecology of an assemblage of large mammals. Hyaenid activity in the lower Pleistocene site at Venta Micena (Orce, Guadix-Baza Basin, Granada, Spain). *Geobios* 31 (3, supplement): 3–47.

Arsuaga, J. L., I. Martinez, C. Lorenzo, A. Gracia, A. Munoz, O. Alonso, and J. Gallego. 1999. The human cranial remains from Gran Dolina Lower Pleistocene site (Sierra de Atapuerca, Spain). *Journal of Human Evolution* 37: 431–57.

Asfaw, B., W. H. Gilbert, Y. Beyene, W. K. Hart, P. R. Renne, G. WoldeGabriel, E. S. Vrba, and T. White. 2002. Remains of *Homo erectus* from Bouri, Middle Awash, Ethiopia. *Nature* 416: 317–20.

Asfaw, B., T. D. White, O. Lovejoy, B. Latimer, S. Simpson, and G. Suwa. 1999. *Australopithecus garhi*: A new species of early hominid from Ethiopia. *Science* 284: 629–35.

Aziz, F. 2002. New discovery of a hominid skull from Cemeng, Sambungmacan, Central Java: an anouncement. *Jurnal Geologi dan Sumberdaya Mineral* 12: 2–7.

Baba, H. 1996. *Reviving* Pithecanthropus. Yomiuri Shinbun, Tokyo.

Baba, H., F. Aziz, Y. Kaifu, G. Suwa, R. T. Kono, and T. Jacob. 2003. *Homo erectus* calvarium from the Pleistocene of Java. *Science* 299: 1384–88.

Bartsiokas, A. 2002. Hominid cranial bone structure: A histological study of Omo 1 specimens from Ethiopia using different microscopic techniques. *Anatomical Record* 267: 52–59.

Bar-Yosef, O. 1998. Early colonizations and cultural continuities in the Lower Paleolithic of Western Asia. In *Early Human Behavior in Global Context: The Rise and Diversity of the Lower Palaeolithic Record*. Edited by M. Petraglia and R. Korisettar. Routledge, London, pp. 221–79.

Bar-Yosef, O., and A. Belfer-Cohen. 2000. Early human dispersals: The unexplored constraint of African diseases. In *Early Humans at the Gates of Europe*. Edited by D. Lordkipanidze, O. Bar-Yosef, and M. Otte. Proceedings of the First International Symposium, Dmanisi, Tbilisi, Georgia. *Études et Recherches Archaeologiques de l'Université de Liège* 92: 79–86.

———. 2001. From Africa to Eurasia—Early dispersals. *Quaternary International* 75: 19–28.

Beeman, C. L. 1959. Peking Man: Letters. *Science* 130: 416.

Bilsborough, A. 2000. Chronology, variability and evolution in *Homo erectus. Variability and Evolution* 8: 5–30.

Binford, L. R., and C. K. Ho. 1985. Taphonomy at a distance: Zhoukoudian, "the cave home of Beijing man?" *Current Anthropology* 26: 413–42.

Binford, L. R., and N. M. Stone. 1986. Zhoukoudian: A closer look. *Current Anthropology* 27: 453–75.

Black, D. 1925. Asia and the dispersal of primates. *Bulletin of the Geological Society of China* 4 (2) 133–83.

———. 1926. Tertiary man in Asia: The Choukoutien discovery. *Nature* 118: 733–34.

———. 1927. The lower molar hominid tooth from the Chou Kou Tien deposit. *Palaeontologia Sinica,* Series D, 7(1): 1–28.

———. 1928. A Study of Kansu and Honan Aeneolithic skulls and specimens from Later Kansu prehistoric sites in comparison with North China and other recent crania. I. On measurement and identification. *Palaeontologia Sinica,* Series D, 6(1): 1–83.

———. 1929. Preliminary note on additional *Sinanthropus* material discovered in Chou Kou Tien in 1928. *Bulletin of the Geological Society of China,* 8: 15–32.

———. 1931a. On the adolescent skull of *Sinanthropus pekinensis* in comparison with an adult skull of the same species and with other hominid skulls, recent and fossil. *Palaeontologia Sinica,* Series D, 7(2): 1–144.

———. 1931b. Evidences of the use of fire by *Sinanthropus*. *Bulletin of the Geological Society of China* 11: 107–98.

Black, D., P. Teilhard de Chardin, C. C. Young, and W. C. Pei. 1933. Fossil Man in China: The Choukoutien Cave Deposits with a Synopsis of Our Present Knowledge of the Late Cenozoic in China. *Memoirs of the Geological Society of China,* Peiping, Series A, No. 11.

Blumenshine, R., J. A. Cavallo, and S. D. Capaldo. 1994. Competition for carcasses and early hominid behavioral ecology: A case study and conceptual framework. *Journal of Human Evolution* 27: 197–213.

Boaz, N. T. 1975. *Interview with G.H.R. von Koenigswald.* Senckenberg Museum, Frankfurt.

———. 1979. Early hominid population densities: New estimates. *Science* 206: 592–95.

———. 1981. History of American paleoanthropological research on early Hominidae, 1925–1980. *American Journal of Physical Anthropology* 56: 397–406.

———. 1997a. *Eco Homo.* Basic Books, New York.

———. 1997b. Calibration and extension of the record of Plio-Pleistocene Hominidae. In *Biological Anthropology: The State of the Science,* 2nd ed. Edited by N. T. Boaz and L. Wolfe. International Institute of Human Evolutionary Research, Bend, Oregon, pp. 25–52.

Boaz, N. T., and A. J. Almquist. 2002. *Biological Anthropology,* 2nd ed. Prentice Hall, Upper Saddle River, N.J.

Boaz, N. T., and R. L. Ciochon. 2001. The scavenging of "Peking Man." *Natural History* 110: 46–51.

Boaz, N. T., and F. C. Howell. 1977. A gracile hominid cranium from upper Member G of the Shungura Formation, Ethiopia. *American Journal Physical Anthropology* 46: 93–108.

Boaz, N. T., D. Ninkovich, and M. Rossignol-Strick. 1982. Paleoclimatic setting for *Homo sapiens neanderthalensis*. *Naturwissenschaften* 69: 29–33.

Boaz, N. T., P. Pavlakis, and A. S. Brooks. 1990. Late Pleistocene-Holocene human remains from the Upper Semliki, Zaire. In *Evolution of Environments and Hominidae in the African Western Rift Valley*. Edited by N. T. Boaz. Virginia Museum of Natural History Memoir, No. 1., pp. 3–14.

Boaz, N. T., R. L. Ciochon, Q. Xu, and J. Liu. 2000. Large mammalian carnivores as a taphonomic factor in the bone accumulation at Zhoukoudian. *Acta Anthropologica Sinica*, Supplement 19: 224–34.

Boesch, C., and H. Boesch. 1990. Tool use and tool making in wild chimpanzees. *Folia Primatologica* 54: 86–99.

Boule, M. 1937. Le Sinanthrope. *L'Anthropologie* 47: 1–22.

Brain, C. 1981. *The Hunters or the Hunted? An Introduction to African Cave Taphonomy*. University of Chicago Press, Chicago.

Bräuer, G. 1994. How different are Asian and African *Homo erectus*? *Courier Forschungs-Institut Senckenberg* 171: 301–18.

Breuil, H. 1931. Le feu et l'industrie lithique et osseuse à Choukoutien. *Bulletin of the Geological Society of China* 11: 147–54.

———. 1939. Bone and antler industry of the Choukoutien *Sinanthropus* site. *Palaeontologia Sinica* New Series D, 6: 1–93.

Brodrick, A. H. 1963. *The Abbé Breuil Prehistorian. A Biography*. Hutchinson, London.

Brown, F. 1994. Development of Pliocene and Pleistocene chronology of the Turkana Basin, East Africa, and its relation to other sites. In *Integrative Paths to the Past: Paleoanthropological Advances in Honor of F. Clark Howell*. Edited by R. S. Corruccini and R. L. Ciochon. Prentice Hall, Englewood Cliffs, New Jersey, pp. 285–312.

Brown, F., J. Harris, R. E. Leakey, and A. Walker. 1985. Early *Homo erectus* skeleton from West Lake Turkana, Kenya. *Nature* 316: 788–92.

Brown, P. 1994. Cranial vault thickness in Asian *Homo erectus* and *Homo sapiens*. *Courier Forschungs-Institut Senckenberg* 17: 33–46.

Cann, R. L., M. Stoneking, and A. C. Wilson. 1987. Mitochondrial DNA and human origins. *Nature* 325: 31–36.

Chaney, R. 1935. The food of "Peking man." *Carnegie Inst. Washington, News Serv. B*, 3(25): 198–202.

Chia, L. (Jia, L.). 1978. A note on the weather conditions in Zhoukoudian area of Peking Man's time. *Acta Stratigraphica Sinica* 2: 53–56.

Chiu, C., Y. M. Ku, Y. Y. Chang, and S. S. Chang. 1973. Peking man fossils and cultural remains newly discovered at Choukoutien. *Vertebrata PalAsiatica* 11: 109–131.

Chow, T., and Y. H. Li. 1960. Report on the excavation of the Choukoutien *Sinanthropus* site in 1959. *Vertebrata PalAsiatica* 2: 97–99.

Ciochon, R. L. 1995. The earliest Asians Yet. *Natural History* 104(12): 50–54.

———. 1996. How it all started? A cave in China sheds new light on man's presence in Asia. *Asiaweek* 22(2): 36–37.

Ciochon, R. L., and J. W. Olsen. 1991. Paleoanthropological and archaeological discoveries from Lang Trang Caves: A new Middle Pleistocene hominid site from northern Vietnam. *Indo-Pacific Prehistory Association Bulletin* 10: 59–73.

Ciochon, R. L., and J. G. Fleagle. 1993. *The Human Evolution Source Book*. Prentice Hall, Englewood Cliffs, New Jersey.

Ciochon, R. L. and R. Larick. 2000. Early *Homo erectus* tools in China. *Archaeology* 53(1): 14–15.

Ciochon, R. L., and K. L. Eaves-Johnson. 2003. China: Archaeological Caves. In *Encyclopedia of Caves and Karst Science*. Edited by J. Gunn. Fitzroy Dearborn, New York and London, pp. 221–25.

Ciochon, R., J. Olsen, and J. James. 1990. *Other Origins: The Search for the Giant Ape in Human Prehistory*. Bantam Doubleday, New York.

Ciochon, R. L., Vu The Long, R. Larick, L. González, R. Grün, J. de Vos, C. Yonge, L. Taylor, H. Yoshida, and M. Reagan. 1996. Dated co-occurrence of *Homo erectus* and *Gigantopithecus* from Tham Khuyen Cave, Vietnam. *Proceedings of the National Academy of Sciences, USA* 93: 3016–20.

Clark, J. D., and J. W. K. Harris. 1986. Fire and its roles in early hominid lifeways. *African Archaeological Review* 3: 3–29.

Coon, C. S. 1962. *The Origin of Races*. Alfred A. Knopf, New York.

Compiling Group of the Atlas. 1980. *Atlas of Primitive Man in China*. Science Press, Beijing.

Cronin, J., N. T. Boaz, C. B. Stringer, and Y. Rak. 1981. Tempo and mode in hominid evolution. *Nature* 292: 113–22.

Dart, R. A. 1925. *Australopithecus africanus*, the man-ape of South Africa. *Nature* 115: 195–99.

DeGusta, D., W. H. Gilbert, and S. P. Turner. 1998. Hypoglossal canal size and hominid speech. *Proceedings of the National Academy of Sciences, USA* 96: 1800–4.

Dinesen, I. (K. Blixen). 1937. *Out of Africa*. Putnam, London.

Djubiantono, T., and F. Sémah. 1993. L'île de Java son peuplement. In *Le Pithecanthrope de Java*. Edited by F. Sémah, A.-M. Sémah, and T. Djubiantono. *Les Dossiers d'Archéologie*, No. 184, pp. 12–19.

Dobzhansky, T. 1937. *Genetics and the Origin of Species*. Columbia University Press, New York.

———. 1942. Races and methods of their study. *Transactions of the New York Academy of Sciences* 4: 115–33.

Duarte, C., J. Mauricio, P. B. Pettitt, P. Souto, E. Trinkaus, H. van der Plicht, and J. Zilhao. 1999. The early Upper Paleolithic human skeleton from Abrigo do Lagar Velho (Portugal) and modern human emergence in Iberia. *Proceedings of the National Academy of Sciences* 96: 7604–9.

Dubois, E. 1894. *Pithecanthropus erectus, eine menschenaehnliche Uebergangsform aus Java.* Landesdruckerei, Batavia, Java.

Elgar, M. A., and B. J. Crespi (editors). 1992. *Cannibalism: Ecology and evolution among diverse taxa.* Oxford University Press, Oxford.

Eller, E. 2002. Population extinction and recolonization in human demographic history. *Mathematical Biosciences* 177/178: 1–10.

Falk, D. 1992. *Braindance.* Henry Holt, New York.

Fernandez, Y., J. C. Diez, I. Caceres, and J. Rosell. 1999. Human cannibalism in the early Pleistocene of Europe (Gran Dolina, Sierra de Atapuerco, Burgos, Spain). *Journal of Human Evolution* 37: 591–622.

Fleagle, J. G. 1998. *Primate Evolution and Adaptation*, 2nd ed. Academic Press, London.

Flint, R. F. 1971. *Glacial and Quaternary Geology.* Wiley, New York.

Fortuyn, A. B. D. 1942. *Excerpt from report of Dr. Fortuyn on Department of Anatomy, P.U.M.C.—Part Two, dated December 5, 1941, with cover memorandum from "AMP" [Agnes M. Pearce, Secretary, China Medical Board, Inc.].* Rockefeller Archives, Sleepy Hollow, N.Y. Record Group 1.1, Series 601D, Box 39, Folder 323.

Gabunia, L., A. Vekua, and D. Lordkipanidze. 2000. The environmental contexts of early human occupation of Georgia (Transcaucasia). *Journal of Human Evolution* 38(2): 785–802.

Gallenkamp, C. 2001. *Dragon Hunter: Roy Chapman Andrews and the Central Asiatic Expeditions.* Viking, New York.

Gibney, F. (editor). 1995. *Senso, The Japanese Remember the Pacific War.* Armonk, N.Y.

Goldberg, P., S. Weiner, O. Bar-Yosef, Q. Wu, and J. Liu. 2001. Site formation processes at Zhoukoudian, China. *Journal of Human Evolution* 41: 483–530.

Goren-Inbar, N., C. S. Feibel, K. L. Verosub, Y. Melamed, M. E. Kislev, E. Tchernov, and I. Saragusti. 2000. Pleistocene milestones on the out-of-Africa corridor at Gesher Benot Ya'aqov, Israel. *Science* 289: 944–47.

Gregory, W. K. 1949. Franz Weidenreich, 1873–1948. *American Anthropologist* 51: 85–90.

Groves, C. P., and V. Mazak. 1975. An approach to the taxonomy of the Hominidae: Gracile Villafranchian hominids of Africa. *Casopis pro Mineralogii a Geologii* 20: 225–47.

Grün, R. 2000. Electron spin resonance dating. In *Modern Analytical Methods in Art and Archaeology.* Edited by E. Ciliberto and G. Spoto. Chemical Analysis Series, Vol. 15. John Wiley, New York, pp. 641–79.

Grün, R., P. H. Huang, X. Wu, C. B. Stringer, A. G. Thorne, and M. McCulloch. 1997. ESR analysis of teeth from the palaeoanthropological site of Zhoukoudian, China. *Journal of Human Evolution* 32: 83–91.

Gue, S., S. Liu, and F. Zhang. 1980. Age determination of Peking Man by fission track dating [In Chinese]. *Kexue Tong Bao* 25: 535.

Haeckel, E. 1868. *Natürliche Schöpfungsgeschichte*. Fischer, Jena.

Hasebe, K. 1915. Notes on the local customs of the Marshall Islands. *Journal of the Anthropological Society of Tokyo* 30(7): 278–79.

———. 1948. A human innominate bone from Lower Pleistocene deposits at Nishiyagi, Akashi, Japan. *Journal of the Anthropological Society of Nippon* 60: 32–36.

Hawks, J. 2003. The browridge: Pleistocene body armor. *American Journal of Physical Anthropology*, Supplement 36: 112.

Hawks, J., K. Hunley, S. Lee, and M. Wolpoff. 2000. Population bottlenecks and Pleistocene human evolution. *Molecular Biological Evolution* 17: 2–22

Heberer, G. 1963. Über einen neuen archanthropinen typus aus der Oldoway-Schlucht. *Zeitschrift für Morphologie und Anthropologie* 53: 171–77.

Heslop, D., M. J. Dekkers, and C. G. Langereis. 2002. Timing and structure of the mid-Pleistocene transition: Records from the loess deposits of northern China. *Palaeogeography, Palaeoclimatology, Palaeoecology* 185: 133–43.

Hoberg, E., N. L. Alkire, A. de Queiroz, and A. Jones. 2001. Out of Africa: Origins of the *Taenia* tapeworms in humans. *Proceedings of the Royal Society of Biological Sciences* 268(1469): 781–87.

Hood, D. 1964. *Davidson Black, A Biography*. University of Toronto Press, Toronto.

Hopwood, D. E. 2003a. Behavioural differences in the early to mid-Pleistocene: Were African and Chinese *Homo erectus* really that different? *American Journal of Physical Anthropology*, Supplement 36: 117–18.

———. 2003b. Examination and interpretation of behavioural change in *Homo erectus* populations from Africa and China. Unpublished master's thesis, Department of Anthropology, Binghamton University, Binghamton, N.Y.

Hou, Y., R. Potts, Y. Baoyin, G. Zhengtang, A. Deino, W. Wei, J. Clark, X. Guangmao, and H. Weiwen. 2000. Mid-Pleistocene Acheulean-like stone technology of the Bose Basin, South China. *Science* 287: 1622–26.

Houghton, H. S. 1941. Letter to Edwin Lobenstine (April 10, 1941). Rockefeller Foundation Archives, Sleepy Hollow, N.Y. Record Group 1.1, Series 601D, Box 39, Folder 323.

Howell, F. C. 1965. *Early Man*. Time-Life, New York.

———. 1999. Paleodemes, species, clades and extinctions in the Pleistocene. *Journal of Anthropological Research* 55: 191–243.

Howells, W. W. 1980. *Homo erectus*—Who, when, and where: A survey. *Yearbook of Physical Anthropology* 23: 1–23.

Hrdlička, A. 1920. Shovel-shaped teeth. *American Journal of Physical Anthropology* 3: 429–65.

Hsu, J. 1965. The climatic conditions in north China during the time of *Sinanthropus* [in Chinese]. *Scientia Sinica* 15: 410–14.

Hu, C. 1990. Letter to Lanpo Jia (March 4, 1977), reprinted in *The Story of Peking Man, from Archaeology to Mystery*. Edited by L. Jia and W. Huang. Oxford University Press, Oxford, pp. 160–61.

Huang, W. 1960. Re-study of the CKT *Sinanthropus* deposits. *Vertebrata PalAsiatica* 2: 83–96.

Huang, W., R. L. Ciochon, Y. Gu, R. Larick, Q. Fang, C. Yonge, J. de Vos, H. Schwarcz, and W. Rink. 1995. Early *Homo* and associated artefacts from Asia. *Nature* 378: 275–78.

Huffman, O. F. 2001. Geologic context and age of the Perning/Mojokerto *Homo erectus*, East Java. *Journal of Human Evolution* 40: 353–62.

Ingersoll, E. 1928. *Dragons and Dragon Lore*. Payson & Clarke, New York.

Isaac, G. L. 1975. Sorting out the muddle in the middle: An anthropologist's post-conference appraisal. In *After the Australopithecines*. Edited by K. Butzer and G. L. Isaac. Mouton, The Hague, pp. 875–87.

Isbell, L. A., J. D. Pruetz, M. Lewis, and T. P. Young. 1998. Locomotor activity differences between sympatric patas monkeys (*Erythrocebus patas*) and vervet monkeys (*Cercopithecus aethiops*): Implications for the evolution of long hindlimb length in *Homo*. *American Journal of Physical Anthropology* 105: 199–207.

Janus, C., and W. Brashler. 1975. *The Search for Peking Man*. Macmillan, New York.

Jia, L. (editor). 1999. *Chronicle of Zhoukoudian (1927–1937)*. Shanghai Scientific and Technical Publishers, Shanghai.

———. 2000. *Chronicle of Zhoukoudian (1927–1937)*. Shanghai Scientific and Technical Publishers, Shanghai.

Jia, L., and W. Huang. 1990. *The Story of Peking Man, from Archaeology to Mystery*. Oxford University Press, Oxford.

Johnsgard, P., and K. Johnsgard. 1982. *Dragons and Unicorns: A Natural History*. St. Martins Press, New York.

Jouy-Avantin, F., A. Debenath, A.-M. Moigne, and H. Moné. 2003. A standardized method for the description and the study of coprolites. *Journal of Archaeological Science* 30(3): 367–72.

Jurmain, R., L. Kilgore, W. Trevathan, and H. Nelson. 2003. *Introduction to Physical Anthropology*, 9th ed. Wadsworth, Belmont, Calif.

Kay, R., M. Cartmill, and M. Balow. 1998. The hypoglossal canal and the origin of human vocal behavior. *Proceedings of the National Academy of Sciences, USA* 95: 5417–19.

Keith, A. 1927. The brain of Anatole France. *British Medical Journal* 2: 1048–49.

Klein, R. 1999. *The Human Career*. University of Chicago Press, Chicago.

Klein, R. G., and B. Edgar. 2002. *The Dawn of Human Culture*. John Wiley & Sons, New York.

Koenigswald, G. H. R. von. 1931a. *Sinanthropus, Pithecanthropus* en de ouderdom van de Trinil-Lagen. *Mijningenieur* 1931: 198–202.

———. 1931b. Fossilen uit Chineesche apotheen in West-Java. *Mijningenier* 1931: 189–93.

———. 1932. Versteinerungen als Azneimittel bei den Chinesen auf Java. *Natur und Museum* 62(9): 292–95.

———. 1936. Erste Mittelung über einen fossilen Hominiden aus dem Altpleistozän Ostjavas. *Natur und Museum* 62: 292–95.

———. 1937. Ein Underkieferfragment des *Pithecanthropus* aus dem Trinilschichten Mittlejavas. *Koninkl. Nederl. Akademie van Wetenschappen* 40: 883–93.

———. 1938. Ein neuer *Pithecanthropus*-Schädel. *Koninkl. Nederl. Akademie van Wetenschappen* 42: 185–92.

———. 1952. *Gigantopithecus blacki* von Koengiswald, a giant fossil hominoid from the Pleistocene of southern China. *Anthropological Papers of the American Museum of Natural History* 43: 291–326.

———. 1956. *Meeting Prehistoric Man*. Harper & Brothers, New York.

———. 1976. *The Evolution of Man*. University of Michigan Press, Ann Arbor.

Koenigswald, G. H. R. von, and F. Weidenreich. 1938. Discovery of an additional *Pithecanthropus* skull. *Nature* 142: 715.

———. 1939. The relationship between *Pithecanthropus* and *Sinanthropus*. *Nature* 144: 926–29.

Krantz, G. S. 1980. Sapienization and speech. *Current Anthropology* 21: 773–92.

Krings, M., H. Geisert, R. W. Schmitz, H. Krainitzki, and S. Paabo. 1999. DNA sequence of the mitochondrial hypervariable region II from the Neandertal type specimen. *Proceedings of the National Academy of Sciences* 96: 5581–85.

Kung, Z. 1981. Discussion on the environmental changes during Peking Man's time and earlier or later than it as viewed from the analysis of pollen [In Chinese]. *Kexue Tong Bao* 25: 1065.

Laitman, J., and R. C. Heimbuch. 1982. The basicranium of Plio-Pleistocene hominids as an indicator of their upper respiratory systems. *American Journal of Physical Anthropology* 59: 323–43.

Larick, R., and R. L. Ciochon. 1996a. The African emergence and early Asian dispersals of the genus *Homo*. *American Scientist* 84: 538–51.

———. 1996b. The first Asians. *Archaeology* 49(1): 51–53.

Larick, R., R. L. Ciochon, and Y. Zaim. 1999. Fossil Farming in Java. *Natural History* 108(6): 54–57.

———. 2004. *Homo erectus*, and the emergence of Sunda in the Tethys Realm. *Athena Review: The Journal of Archaeology, History, and Exploration* 4(1): 32–39.

Larick, R., R. L. Ciochon, Y. Zaim, Sudijono, Suminto, Y. Rizal, F. Aziz, J. Arif, M. Reagan, and M. Heizler. 2001. Early Pleistocene ^{40}Ar/^{39}Ar ages for Bapang Formation hominins, Central Java, Indonesia. *Proceedings of the National Academy of Sciences, USA* 98: 4866–71.

Leakey, L. 1961. New finds at Olduvai Gorge. *Nature* 189: 649–50.

Leakey, L., P. V. Tobias, and J. R. Napier. 1964. A new species of the genus *Homo* from Olduvai Gorge. *Nature* 202: 7–9.

LeCount, E. R., and C. W. Apfelbach. 1920. Pathologic anatomy of traumatic fractures of the cranial bones and concomitant brain injuries. *Journal of the American Medical Association* 74: 501–11.

Le Fort, R. 1901. Étude experimentale sur les fractures de la machoire supérieure. *Revue de Chirurgie* 23: 201–10.

Lewin, R. 1989. *Bones of Contention*. Simon & Schuster, New York.

Li, M. S., and N. Yue. 2000. *Search for Peking Man* [In Chinese]. Hua Xia Publishers, Beijing.

Lin, J. 1999. Mystery of the Missing Ancient Skulls. *Detroit Free Press*, May 17, 1999. (http://www.100megsfree4.com/farshores/skulls.htm).

Lin, S. 1994. *Zhoukoudian Peking Man Site: A World's Significant Site of Cultural Importance*. Institute of Vertebrate Paleontology and Paleoanthropology, and Chinese Academy of Sciences, Beijing.

Liu, Z. 1985. Sequence of sediments at Locality 1 in Zhoukoudian and correlation with loess stratigraphy in northern China and with the chronology of deep-sea cores. *Quaternary Research* 23: 139–53.

MacLarnon, A. 1993. The vertebral canal. In *The Nariokotome* Homo erectus *Skeleton*. Edited by A. Walker and R. Leakey. Harvard University Press, Cambridge, pp. 359–90.

Matthew, W. D. 1939. *Climate and Evolution*. Academy of Natural Sciences, New York.

Mayr, E. 1950. Taxonomic categories in fossil hominids. In *Origins and Evolution of Man*. Cold Springs Harbor Symposium on Quantitative Biology. Vol. 15. The Biological Laboratory, Cold Springs Harbor, N.Y., pp. 109–18.

McCown, T. D., and A. Keith. 1937–39. *The Stone Age of Mount Carmel. II. The Fossil Human Remains from the Levalloiso-Mousterian*. Clarendon, Oxford.

Milius, S. 2001. Tapeworms tell tales of deeper human past. *Science News* 159: 215–6.

Moore, R. 1953. *Men, Time and Fossils: The Story of Evolution.* Alfred A. Knopf, New York.

Mortier, J., and M. L. Auboux. 1966. *Teilhard de Chardin Album.* Harper & Row, New York.

Movius, H. L. 1969. Lower Paleolithic archaeology in southern Asia and the Far East. In *Studies in Physical Anthropology: Early Man in the Far East.* Edited by W. W. Howells. Anthropological Publications, Netherlands, pp. 17–77.

O'Connell, J., K. Hawkes, and N. G. Blurton Jones. 1999. Grandmothering and the evolution of *Homo erectus. Journal of Human Evolution* 36: 461–86.

Online Mendelian Inheritance in Man. http://www.ncbi.nlm.nih.gov (accessed July, 2003).

Opitz, J. M., and M. C. Holt. 1990. Microcephaly: General considerations and aids to nosology. *Journal of Craniofacial Genetics and Developmental Biology* 10(2): 175–204.

Osborn, H. F. 1922. *Hesperopithecus,* the first anthropoid primate found in America. *American Museum Novitates,* No. 37: 1–5.

———. 1924. American men of the dragon bones. *Natural History* 24(3): 350–65.

Oosterzee, P. van. 2000. *Dragon Bones: The Story of Peking Man.* Perseus, Cambridge, Massachusetts.

Pei, J., and J. Sun. 1979. Thermoluminescence ages of quartz in ash materials from *Homo erectus pekinensis* site and its geological implication [In Chinese]. *Kexue Tong Bao* 24: 849.

Pei, W. 1929. An account of the discovery of an adult *Sinanthropus* skull in the Choukoutien cave deposit. *Bulletin of the Geological Society of China* 8: 203–5.

———. 1931a. Notice of the discovery of quartz and other stone artifacts in the Lower Pleistocene hominid-bearing sediments of the Choukoutien cave deposit. *Bulletin of the Geological Society of China* 11: 109–46.

———. 1931b. Mammalia remains from Locality 5 at Choukoutien. *Palaeontologia Sinica,* Series C, 7(2): 1–18.

———. 1931c. The age of Choukoutien fossiliferous deposits. *Bulletin of the Geological Society of China* 10: 165–78.

———. 1933. A preliminary report on the Late-Paleolithic cave of Choukoutien. *Bulletin of the Geological Society of China* 13: 327–58.

———. 1936. On the mammalian remains from Locality 3 at Choukoutien. *Palaeontologia Sinica,* Series C, 7: 1–120.

———. 1937a. Histoire des découvertes et organisation des fouilles. *Bulletin de la Société Préhistorique Française* 34: 354–66.

———. 1937b. Le rôle des phénomènes naturels dans l'éclatement et le façonnement des roches dures utilisées par l'homme prehistorique. *Revue Géologique Physique et Géologique Dynamique* 9: 1–78.

———. 1937c. Les fouilles de Choukoutien en Chine. *Bulletin de la Société Préhistorique Française* 34: 353–66.

———. 1939. A preliminary study on a new palaeolithic station known as Locality 15 within the Choukoutien region. *Bulletin of the Geological Society of China* 19: 147–87.

Pei, W., and S. Zhang. 1985. *A Study of the Lithic Artifacts of* Sinanthropus. *Palaeontologia Sinica,* New Series D, 12: 1–277. Science Press, Beijing.

Pope, G. G. 1983. Evidence on the age of the Asian Hominidae. *Proceedings of the National Academy of Sciences, USA* 80: 4988–92.

———. 1988a. Current issues in Far Eastern palaeoanthropology. In *The Palaeoenvironment of East Asia from the Mid-Tertiary.* Vol. II. Edited by P. Whyte et al. University of Hong Kong Centre of Asian Studies, Hong Kong, pp. 1097–1123.

———. 1988b. Recent advances in Far Eastern paleoanthropology. *Annual Review of Anthropology* 17: 43–77.

———. 1989. Bamboo and Human Evolution. *Natural History* 10: 49–57.

———. 1993. Ancient Asia's cutting edge. *Natural History* 5: 55–59.

Potts, R. 1996. *Humanity's Descent: The Consequences of Ecological Instability.* Morrow, New York.

Proctor, R. 1988. From *Anthropologie* to *Rassenkunde* in the German anthropological tradition. In *Bones, Bodies, Behavior. Essays on Biological Anthropology.* History of Anthropology, Vol. 5. Edited by G. W. Stocking, Jr. University of Wisconsin Press, Madison, pp. 138–79.

Provine, W. B. 1986. *Sewell Wright and Evolutionary Biology.* University of Chicago Press, Chicago.

Qian, F., J. Zhang, and J. Li. 1980. Magnetostratigraphic study of the cave deposit containing fossil Peking Man at Zhoukoudian [In Chinese]. *Kexue Tong Bao* 25: 359.

Rasky, H. (editor). 1977. *The Peking Man Mystery.* Canadian Broadcasting Company, Toronto.

Reader, J. 1981. *Missing Links, The Hunt for Earliest Man.* Little, Brown, Boston.

Rightmire, G. P. 1990. *The Evolution of* Homo erectus: *Comparative Anatomical Studies of an Extinct Human Species.* Cambridge University Press, Cambridge.

———. 1998. Human evolution in the Middle Pleistocene: The role of *Homo heidelbergensis. Evolutionary Anthropology* 6: 218–27.

———. 2001. Patterns of hominid evolution and dispersal in the Middle Pleistocene. *Quaternary International* 75: 77–84.

Rogers, L. F. 1992. *Radiology of Skeletal Trauma.* Churchill Livingstone, New York.

Rothschild, B. M., I. Hershkovitz, and C. Rothschild. 1995. Origin of yaws in the Pleistocene. *Nature* 378: 343.

Rowlett, R. 1999. Did the use of fire for cooking lead to a diet change that resulted in the expansion of brain size in *Homo erectus* from that of *Australopithecus africanus? Science* 283: 2005.

Ruff, C. 1991. Climate and body shape in hominid evolution. *Journal of Human Evolution* 21: 81–105.

Sartono, S. 1971. Observations on a new skull of *Pithecanthropus* erectus (*Pithecanthropus* VIII) from Sangiran, Central Java. *Courier Forschungs-Institut Senckenberg* 74(2): 185–94.

Saunders, J. J., and B. K. Dawson. 1998. Bone damage patterns produced by extinct hyena, *Pachycrocuta brevirostris* (Mammalia: Carnivora), at the Haro River Quarry, Northwestern Pakistan. In *Advances in Vertebrate Paleontology and Geochronology.* Edited by Y. Tomida, L. J. Flynn, and L. L. Jacobs. National Science Museum Monographs, No. 14, Tokyo, pp. 215–42.

Savage, R. J. G., and M. R. Long. 1986. *Mammalian Evolution: An Illustrated Guide.* Facts on File, New York.

Schick, K. D., and N. S. Toth. 1993. *Making Silent Stones Speak: Human Evolution and the Dawn of Technology.* Simon & Schuster, New York.

Schick, K. D., N. S. Toth, W. Qi, J. D. Clark, and D. Etler. 1991. Archaeological perspectives in the Nihewan Basin, China. *Journal of Human Evolution* 27: 13–26.

Schlosser, M. 1903. Die fossilen Säugethiere Chinas. *Abhandlungen der Bayerischen Akademie der Wissenschaften* II, 150: 22.

Semaw, S., P. Renne, J. W. K. Harris, C. S. Feibel, R. L. Bernor, N. Fesseha, and K. Mowbray. 1997. 2.5-million-year-old stone tools from Gona, Ethiopia. *Nature* 385: 333–36.

Semaw, S. 2000. The world's oldest stone artifacts from Gona, Ethiopia: Their implications for understanding stone technology and patterns of human evolution between 2.6–1.5 million years ago. *Journal of Archaeological Science* 27: 1197–214.

Shapiro, H. 1974. *Peking Man: The Discovery, Disappearance and Mystery of a Priceless Scientific Treasure.* Simon & Schuster, New York.

Shen, G., and L. Jin. 1993. Restudy of the upper age limit of Beijing man site. *International Journal of Anthropology* 8: 95–98.

Shen, G., T. L. Ku, H. Cheng, R. L. Edwards, Z. Yuan, and Q. Wang. 2001. High-precision U-series dating of Locality 1 at Zhoukoudian, China. *Journal of Human Evolution* 41: 679–88.

Shen, G., W. Wang, Q. Wang, J. Zhao, K. Collerson, C. Zhou, and P. V. Tobias. 2002. U-series dating of Liujiang hominid site in Guangxi, southern China. *Journal of Human Evolution* 43(6): 817–29.

Shen, L., S. Zhang, and J. Li. 1981. Mineral composition of clastic cave deposits from the site of Peking Man and its implications [In Chinese]. *Scientia Geologica Sinica* 1: 60–72.

Shipman, P. 2001. *The Man Who Found the Missing Link: Eugene Dubois and His Lifelong Quest to Prove Darwin Right.* Simon & Schuster, New York.

Sigmon, B., and J. S. Cybulski (editors). 1981. Homo erectus: *Papers in Honor of Davidson Black.* University of Toronto Press, Toronto.

Sillen, A., G. Hall, and R. Armstrong. 1998. $^{87}Sr/^{86}Sr$ ratios in modern and fossil food-webs of the Sterkfontein Valley: Implications for early hominid habitat preferences. *Geochimica et Cosmochimica Acta* 62: 2463–78.

Simpson, G. G. 1945. The principles of classification and the classification of mammals. *Bulletin of the American Museum of Natural History* 85: 1–350.

———. 1964. *Tempo and Mode in Evolution.* Hafner, New York.

Smith, G. E. 1932. The discovery of primitive man in China. In *Smithsonian Report for 1931.* U. S. Government Printing Office, Washington, D.C., pp. 531–47.

Smith, F., E. Trinkaus, P. B. Pettitt, I. Karavanic, and M. Paunovic. 1999. Direct radio-carbon dates for Vindija G_1 and Velika Pecina Late Pleistocene hominid remains. *Proceedings of the National Academy of Sciences* 96(22): 12281–86.

Spencer, F. 1990. *Piltdown: A Scientific Forgery.* Oxford University Press, New York.

Spencer, L. M. 1997. Dietary adaptations of Plio-Pleistocene Bovidae: Implications for hominid habitat use. *Journal of Human Evolution* 32: 201–28.

Stanley, S. M. 1998. *Children of the Ice Age: How a Global Catastrophe Allowed Humans to Evolve.* W. H. Freeman, New York.

Stringer, C. B. 1984. The definition of *Homo erectus* and the existence of the species in Africa and Europe. *Courier Forschungs-Institut Senckenberg* 69: 131–43.

Stringer, C. B. and R. McKie. 1997. *African Exodus: The Origins of Modern Humanity.* Henry Holt, New York.

Strum, S. C., D. Lindburg, and D. Hamburg (editors). 1999. *The New Physical Anthropology: Science, Humanism, and Critical Reflection.* Prentice Hall, Upper Saddle River, New Jersey.

Swisher III, C. C., G. H. Curtis, and R. Lewin. 2000. *Java Man.* Scribner, New York.

Swisher III, C. C., G. H. Curtis, T. Jacob, A. G. Getty, A. Suprijo, and Widiasmoro. 1994. Age of the earliest known hominids in Java, Indonesia. *Science* 266: 1118–21.

Swisher III, C. C., W. J. Rink, S. C. Anton, H. P. Schwarcz, G. H. Curtis, A. Suprijo, and Widiasmoro. 1996. Latest *Homo erectus* in Java: Potential contemporaneity with *Homo sapiens* in southeast Asia. *Science* 274: 1870–74.

Tan, A. 2001. *The Bonesetter's Daughter.* Ballantine Books, New York.

Taplin, A. 1874. *The Narrinyeri: An Account of the Tribes of South Australian Aborigines Inhabiting the Country Around the Lakes Alexandrina, Albert, and Coorong, and the Lower Part of the River Murray*. E. S. Wigg and Son, Adelaide.

Taschdjian, C. 1977. *The Peking Man is Missing*. Harper & Row, New York.

Tattersall, I. 1995. *The Fossil Trail*. Oxford University Press, Oxford.

Tattersall, I. and J. H. Schwartz. 1999. Hominids and hybrids: The place of Neandertals in human evolution. *Proceedings of the National Academy of Sciences* 96: 7117–19.

Teilhard de Chardin, P. 1934. Letter to Walter Granger (March 19, 1934). Granger-Teilhard Collection, Georgetown University Library, letter 1: 11.

———. 1935a. Letter to Walter Granger (March 27, 1935). Granger-Teilhard Collection, Georgetown University Library, letter 1: 17.

———. 1935b. Letter to Walter Granger (July 25, 1935). Granger-Teilhard Collection, Georgetown University Library, letter 1: 13.

———. 1936. Letter to Walter Granger (February 18, 1936). Granger-Teilhard Collection, Georgetown University Library.

———. 1941. *Early Man in China*, No. 7. Institut de Géo-Biologie, Pékin.

Teilhard de Chardin, P. and W. Pei. 1932. The lithic industry of the *Sinanthropus* deposits in Choukoutien. *Bulletin of the Geological Society of China* 11: 317–58.

———. 1933. New discoveries in Choukoutien 1933–1934. *Bulletin of the Geological Society of China* 13: 369–94.

Teilhard de Chardin, P. and C. C. Young. 1930. Preliminary report on the Choukoutien fossiliferous deposit. *Bulletin of the Geological Society of China* 8: 173–202.

———. 1932. Fossil mammals from the Late Cenozoic of North China. *Palaeontologia Sinica*, Series C, 9: 1–84.

Templeton, A. 2002. Out of Africa again and again. *Nature* 416: 45–51.

Tishkoff, S. A., R. Varkonyi, N. Cahinhinan, S. Abbes, G. Argyropoulos, G. Destro-Bisol, A. Drousiotou, B. Dangerfield, G. Lefranc, J. Loiselet, A. Piro, M. Stoneking, A. Tagarelli, G. Tagarelli, E. Touma, S. Williams, and A. Clark. 2001. Haplotype diversity and linkage disequilibrium at human *G6PD*: Recent origin of alleles that confer malarial resistance. *Science* 293: 455–62.

Tobias, P. V. 1976. The life and times of Ralph von Koenigswald: paleontologist extraordinary. *Journal of Human Evolution* 5: 403–12.

———. 1991. *Olduvai Gorge: Vol. 4. The Skulls, Endocasts, and Teeth of* Homo habilis. Cambridge University Press, Cambridge.

Tobias, P., W. Qian, and J. L. Cormack. 2000. Davidson Black and Raymond Dart: Asian-African parallels in palaeoanthropology. *Acta Anthropologica Sinica*, Supplement 19: 299–306.

Tobias, P., and G. H. R. von Koenigswald. 1964. A comparison between the Olduvai hominines and those of Java and some implications for hominid phylogeny. *Nature* 204: 515–18.

Tong, Y., et al. (editors). 1997. *Evidence for Evolution—Essays in Honor of Prof. Chungchien Young on the Hundredth Anniversary of His Birth*. China Ocean Press, Beijing.

Turner, C. 2002. When People Fled Hyenas: Oversized Hyenas May Have Delayed Human Arrival in North America (review by Lee Dye). http://abcnews.go.com/sections/scitech/DyeHard/dyehard021120.html (accessed June, 2003).

U.S. Department of Health and Human Services. 2002. Deaths: leading causes for 2000. *National Vital Statistics Reports* 50 (16): 1–86.

Villa, P. 1992. Cannibalism in Prehistoric Europe. *Evolutionary Anthropology* 3: 93–104.

Villa, P., C. Bouville, J. Courtin, D. Helmer, E. Mahieu, P. Shipman, G. Belluomini, and M. Branca. 1986. Cannibalism in the Neolithic. *Science* 233: 431–37.

Voris, H. 2000. Maps of Pleistocene sea levels in Southeast Asia: shorelines, river systems and time durations. *Journal of Biogeography* 27: 1153–67.

Walker, A., M. R. Zimmerman, and R. E. Leakey. 1982. A possible case of hypervitaminosis A in *Homo erectus*. *Nature* 296: 248–50.

Walker, A. and R. E. Leakey. 1993. *The Nariokotome* Homo erectus *Skeleton*. Harvard University Press, Cambridge.

Walker, A. and P. Shipman. 1996. *The Wisdom of the Bones: In Search of Human Origins*. Alfred A. Knopf, New York.

Wang, H., S. H. Ambrose, C. Liu, and L. Follmer. 1997. Paleosol stable isotope evidence for early hominid occupation of East Asian temperate environments. *Quaternary Research* 48: 228–38.

Wang, Q., and P. V. Tobias. 2000. Review of the phylogenetic position of Chinese *Homo erectus* in light of midfacial morphology. *Acta Anthropologia Sinica*, Supplement 19: 23–33.

Washburn, S. L. 1946. The effect of facial paralysis on the growth of the skull of rat and rabbit. *Anatomical Record* 94: 163–68.

———. 1947. The relation of the temporal bone to the form of the skull. *Anatomical Record* 99: 239–48.

———. 1951. The new physical anthropology. *Transactions of the New York Academy of Sciences* 13: 298–304.

———. 1964. The origin of races: Weidenreich's opinion. *American Anthropologist* 66: 1165–67.

———. 1983. Evolution of a teacher. *Annual Review of Anthropology* 12: 1–24. Reprinted in *The New Physical Anthropology: Science, Humanism, and Critical Reflection*. Edited by Strum, Shirley, Donald G. Lindburg, and David Hamburg. Prentice Hall, Upper Saddle River, New Jersey, pp. 215–27.

Washburn, S. L., and R. Moore. 1974. *Ape into Man: A Study of Human Evolution*. Little, Brown, Boston.

Washburn, S. L. and D. Wolffson (editors). 1949. *The Shorter Anthropological Papers of Franz Weidenreich Published in the Period 1939–1948: A Memorial Volume*. Viking Fund, New York.

Weaver, W. 1941a. Internal Report (June 6, 1941). Rockefeller Foundation Archives, Sleepy Hollow, N.Y. Record Group 1.1, Series 601D, Box 39, Folder 323.

———. 1941b. Notes on Interview by "WW" with Dr. F. Weidenreich (Friday, June 6, 1941). Rockefeller Foundation Archives, Sleepy Hollow, N.Y. Record Group 1.1, Series 601D, Box 39, Folder 323.

Weaver, W. and R. B. Fosdick. 1941. Notes on Interview by "WW" with "RBF" (Friday, June 6, 1941). Rockefeller Foundation Archives, Sleepy Hollow, NY. Record Group 1.1, Series 601D, Box 39, Folder 323.

Weidenreich, F. 1930. Ein neuer *Pithecanthropus*-Fund in China. *Natur und Museum* 60(12): 546–51.

———. 1932. Ueber pithecoide Merkmale bei *Sinanthropus pekinensis* und seine stammesgeschichte Beurteilung. *Zeitschrift für Anatomie und Entwicklungsgechichte* 99: 212–52.

———. 1935. The *Sinanthropus* population of Choukoutien (Locality 1) with a preliminary report on new discoveries. *Bulletin of the Geological Society of China* 14: 427–61.

———. 1936a. Observations on the form and the proportions of the endocranial casts of *Sinanthropus pekinensis*, other hominids, and the great apes: A comparative study of brain size. *Palaeontologia Sinica*, New Series D, 7(4): 1–50.

———. 1936b. The mandibles of *Sinanthropus pekinensis*: A comparative study. *Palaeontologia Sinica*, New Series D, 7(3): 1–162.

——— 1937. The dentition of *Sinanthropus pekinensis*: A comparative odontography of the hominids. *Palaeontologia Sinica*, New Series D, 1: 1–181 (text); 1–121 (atlas).

———. 1938. The ramification of the middle meningeal artery in fossil hominids and its bearing upon phylogenetic problems. *Palaeontologia Sinica*, New Series D, 3: 1–16.

———. 1939a. *Sinanthropus* and his significance for the problem of human evolution. *Bulletin of the Geological Society of China* 19: 1–17.

———. 1939b. The classification of fossil hominids and their relations to each other, with special reference to *Sinanthropus pekinensis*. *Bulletin of the Geological Society of China* 19: 64–75.

———. 1940. Man or ape? *Natural History* 45: 32–37.

———. 1941a. The brain and its role in the phylogenetic transformation of the skull. *Transactions of the American Philosophical Society* 31: 321–442.

————. 1941b. The extremity bones of *Sinanthropus pekinensis*. *Palaeontologia Sinica*, New Series D, 5: 1–151.

————. 1944. Giant early man from Java and South China. *Science* 99: 479–82.

————. 1945. Giant early man from Java and south China. *Anthropological Papers of the American Museum of Natural History*, 40(1): 1–134.

————. 1946. *Apes, Giants, and Man*. University of Chicago Press, Chicago.

————. 1951. Morphology of Solo Man. *Anthropological Papers of the American Museum of Natural History* 43: 205–90.

Weiner, S., Q. Q. Xu, P. Goldberg, J. Y. Liu, and O. Bar-Yosef. 1998. Evidence for the use of fire at Zhoukoudian, China. *Science* 281: 251–53.

White, T. D., B. Asfaw, D. DeGusta, H. Gilbert, G. D. Richards, G. Suwa, and F. C. Howell. 2003. Pleistocene *Homo sapiens* from Middle Awash, Ethiopia. *Nature* 423: 742–47.

White, T. 1992. *Prehistoric Cannibalism at Mancos 5MTUMR-2346*. Princeton University Press, Princeton.

Williamson, P. G. 1981. Paleontological documentation of speciation in Cenozoic mollusks from Turkana Basin. *Nature* 293: 437–43.

Wolpoff, M. 1980. *Paleoanthropology*. Alfred A. Knopf, New York.

————. 1999. *Paleoanthropology*. 2nd ed. McGraw-Hill, New York.

————. 1984. Evolution in *Homo erectus*: The question of stasis. *Paleobiology* 10: 389–406.

Wolpoff, M. and R. Caspari. 1997. *Race and Human Evolution*. Simon and Schuster, New York.

Wong, W. (Weng, W.) 1927. The search for early man in China. *Bulletin of the Geolological Society of China* 6: 335–36.

Wong, W. H. (Weng, W. H.), and T. H. Yin. 1941. Letter to Dr. H. S. Houghton (January 10, 1941). Rockefeller Archives, Sleepy Hollow, N.Y. Record Group 1.1, Series 601D, Box 39, Folder 323.

Woo, J., and T. K. Chao. 1959. New discovery of *Sinanthropus* mandible from Choukoutien. *Vertebrata PalAsiatica* 4: 17–26.

Wrangham, R., J. H. Jones, G. Laden, D. Pilbeam, and N. Conklin-Brittain. 1999. The raw and the stolen: Cooling and the ecology of human origins. *Current Anthropology* 5: 567–94.

Wright, S. 1940. Breeding structure of populations in relation to speciation. *American Naturalist* 74: 232–48.

————. 1968–1977. *Evolution and the Genetics of Populations. A Treatise*. University of Chicago Press, Chicago.

Wu, R. 1985. Chinese *Homo erectus* and recent work at Zhoukoudian. In *Ancestors: The Hard Evidence*. Edited by E. Delson. Liss, New York, pp. 206–14.

Wu, R., M. E. Ren, X. M. Xu, Z. G. Yang, C. K. Hu, Z. C. Kong, Y. Y. Xie, and S. S. Zhao (editors). 1985. *Multi-disciplinary Study of the Peking Man Site at Zhoukoudian*. Science Press, Beijing.

Wu, R. and S. Lin. 1983. Peking Man. *Scientific American* 248: 86–94.

Wu, R. and J. W. Olsen (editors). 1985. *Palaeoanthropology and Palaeolithic Archaeology in the People's Republic of China*. Academic Press, Orlando, Fla.

Wu, X. 1989. Early *Homo sapiens* in China. In *Early Humankind in China*. Edited by X. Wu and S. Shang, Science Press, Beijing.

Wu, X., and F. E. Poirier. 1995. *Human Evolution in China: A Metric Description of the Fossils and a Review of the Sites*. Oxford University Press, New York.

Wynn, T. 1993. Two developments in the mind of *Homo erectus*. *Journal of Anthropological Archaeology* 12: 299–322.

Xie, Y. 1985. The sedimentary environment of the Peking Man period. In *Multi-disciplinary Study of the Peking Man Site at Zhoukoudian*. Edited by R. Wu, M. E. Ren, X. M. Xu, Z. G. Yang, C. K. Hu, Z. C. Kong, Y. Y. Xie, and S. S. Zhao. Science Press, Beijing, pp. 185–215.

Yuan, S. X., T. M. Chen, S. J. Gao, and Y. Q. Hu. 1991. Study on uranium series dating of fossil bones from Zhoukoudian. *Acta Anthropologica Sinica* 10: 189–93.

Zdansky, O. 1927. Preliminary notice on two teeth of a hominid from a cave in Chihli (China). *Bulletin of the Geological Society of China* 5: 281–84.

———. 1928. Die Säugetiere der Quatärfauna von Choukoutien. *Palaeontologia Sinica*, Series C, 5: 1–146.

Zhao, S., M. Xia, and S. Wan. 1980. Uranium-series dating of Peking Man [In Chinese]. *Kexue Tong Bao* 25: 447.

Zhao, S. 1985. Uranium-series dating of Peking Man site. In *Multi-disciplinary Study of the Peking Man Site at Zhoukoudian*. Edited by R. Wu, M. E. Ren, X. M. Xu, Z. G. Yang, C. K. Hu, Z. C. Kong, Y. Y. Xie, and S. S. Zhao. Science Press, Beijing, pp. 246–55.

Zhou, C., Z. Liu, Y. Wang, and Q. Huang. 2000. Climatic cycles investigated by sediment analysis in Peking Man's cave, Zhoukoudian, China. *Journal of Archaeological Science* 27: 101–9.

Zhou, M., and C. K. Ho. 1990. History of the dating of *Homo erectus* at Zhoukoudian. *Geological Society of America Special Paper* 242: 69–74.

Zhu, R., Y. X. Pan, B. Guo, C. D. Shi, Z. T. Guo, B. Y. Yuan, Y. M. Hou, W. W. Huang, K. A. Hoffman, R. Potts, and C. L. Deng. 2001. Earliest presence of humans in northeast Asia. *Nature* 413: 413–17.

Zuckerkandl, E. and L. Pauling. 1965. Molecules as documents of evolutionary history. *Journal of Theoretical Biology* 8(2): 357–66.

Credits for Illustrations

Black and White Photographs and Line Drawings:

Page v: Courtesy of the Institute of Vertebrate Paleontology and Paleoanthropology, Chinese Academy of Sciences.

Page 2: (top, middle) Modified after the original sketches made by George Barbour about 1929 (from Andersson, 1943, figures 4 and 5); *(bottom)* Modified after figure 2 in Goldberg et al. (2001), courtesy of Paul Goldberg. Digital images created and modified by Michael Zimmerman.

Page 5: Courtesy of Zhoukoudian Museum, Institute of Vertebrate Paleontology and Paleoanthropology, Chinese Academy of Sciences.

Page 11: Photograph by John Reader, from his book *Missing Links* (Reader, 1981, p. 106). Courtesy of John Reader. Digital image created and modified by Michael Zimmerman.

Page 15: Courtesy of Éditions du Seuil, Paris. Digital image created and modified by Michael Zimmerman.

Page 20: Digital image created and modified by Michael Zimmerman.

Page 23: Courtesy of the Institute of Vertebrate Paleontology and Paleoanthropology, Chinese Academy of Sciences.

Page 28: Courtesy of the Institute of Vertebrate Paleontology and Paleoanthropology, Chinese Academy of Sciences.

Page 30: Courtesy of Zhoukoudian Museum, Institute of Vertebrate Paleontology and Paleoanthropology, Chinese Academy of Sciences.

Page 31: (top) Image no. 338922, courtesy of the American Museum of Natural History Library, New York; *(bottom)* Digital photograph by Michael Zimmerman of an American Museum of Natural History cast.

Page 34: Image no. 335797, courtesy of the American Museum of Natural History Library, New York.

Page 37: Courtesy of Éditions du Seuil, Paris. Digital image created and modified by Michael Zimmerman.

Page 41: Courtesy of Éditions du Seuil, Paris. Digital image created and modified by Michael Zimmerman.

Page 44: Courtesy of the Institute of Vertebrate Paleontology and Paleoanthropology, Chinese Academy of Sciences.

Page 52: Image no. 335658, courtesy of the American Museum of Natural History Library, New York.

Page 56: From figure 11-4 in *Introduction to Physical Anthropology* (with InfoTrac) 9th edition by Jurmain / Kilgore / Trevathan / Nelson © 2003. Reprinted with permission of Wadsworth, a division of Thomson Learning: www.thomsonrights.com. fax: 800 730-2215. Digital image modified by Michael Zimmerman.

Page 64: Photograph courtesy of David Gantt taken of the original specimens housed at Senckenberg Museum, Frankfurt, Germany.

Page 66: Redrawn from page 201 in *Race and Human Evolution* (1997) by Milford Wolpoff and Rachel Caspari. Courtesy of Milford Wolpoff. Digital image created and modified by Michael Zimmerman.

Page 69: Image courtesy of Ian Tattersall and Ken Mowbray, American Museum of Natural History. Drawing by Don McGranaghan.

Page 75: Reproduced from plate 6 in Elliot Smith (1932) *Smithsonian Report for 1931.* Courtesy of the Smithsonian Institution, Washington, D.C.

Page 78: From figure 10-18 in *Biological Anthropology: A Synthetic Approach to Human Evolution* 2th edition by Boaz / Almquist © 2002. Reprinted with permission of Prentice Hall, a division of Pearson Education: http://www.prenhall.com. Digital image modified by Michael Zimmerman.

Page 79: Modified after figure on page 45 in Baba (1996) *Reviving* Pithecanthropus. Courtesy of M. Baba. Digital image and labeling by Michael Zimmerman.

Page 82: Reproduced from plate 8 in Elliot Smith (1932) *Smithsonian Report for 1931.* Courtesy of the Smithsonian Institution, Washington, D.C.

Page 87: (top) Photograph by Noel Boaz; *(middle, bottom)* Photographs by Russell Ciochon.

Page 91: Courtesy of the Institute of Vertebrate Paleontology and Paleoanthropology, Chinese Academy of Sciences.

Page 92: Courtesy of the Institute of Vertebrate Paleontology and Paleoanthropology, Chinese Academy of Sciences.

Page 93: (top) Image no. 335795, courtesy of the American Museum of Natural History Library, New York; *(bottom)* Digital image by Nathan Totten.

Page 94: Digital image created by Nathan Totten, refined under the direction of Noel Boaz and Russell Ciochon.

Page 96: (top) Courtesy of Zhoukoudian Museum, Institute of Vertebrate Paleontology and Paleoanthropology, Chinese Academy of Sciences.

Page 96: (bottom) Reproduced from figure 27 in Teilhard de Chardin (1941) *Early Man in China.* Courtesy of the Institut de Géo-Biologie, Pékin.

Page 99: Courtesy of Éditions du Seuil, Paris. Digital image by Michael Zimmerman.

Page 110: (top) Redrawn from figure 23 in Teilhard de Chardin (1941) *Early Man in China.* Courtesy of the Institut de Géo-Biologie, Pékin; *(bottom)* Courtesy of the Institute of Vertebrate Paleontology and Paleoanthropology, Chinese Academy of Sciences.

Page 111: Redrawn from map on page 120 in Jia and Huang (1990) *The Story of Peking Man, from Archaeology to Mystery.* Digital image scanned and modified by Michael Zimmerman and Nathan Totten.

Page 113: Redrawn from figure 20 in Teilhard de Chardin (1941) *Early Man in China.* Courtesy of the Institut de Géo-Biologie, Pékin. Digital image scanned and modified by Michael Zimmerman.

Page 117: Redrawn from and modified after figure 2.10 in Klein (2000) *The Human Career.* Digital image by Erin Schembari.

Page 118: Adapted and redrawn from figure 4 in Zhou et al. (2000) *Journal of Archaeological Science,* volume 27. Digital image by Michael Zimmerman.

Page *121:* Photograph provided by Alan Walker of Pennsylvania State University and reproduced with the permission of the National Musuems of Kenya.

Page 126: Redrawn from figure 6.7 in Washburn and Moore (1974) *Ape into Man.* Digital image by Michael Zimmerman.

Page 131: (top, bottom) Courtesy of Steve Weiner, Weizmann Institute of Science, Rehovat, Israel. Digital image modified by Nathan Totten.

Page 133: Photograph by Russell Ciochon. Digital image by Erin Schembari.

Page 134: (top, middle) Photographs by Russell Ciochon; *(bottom)* Photograph by Noel Boaz. Digital image by Michael Zimmerman.

Page *136: (top, bottom)* Courtesy of Kathy Schick and Nicholas Toth, CRAFT Research Center, Indiana University.

Page 137: (top, bottom): Photographs by Russell Ciochon and Noel Boaz. Digital images and labeling by Michael Zimmerman.

Page 139: Photograph by Noel Boaz and Chris Davett.

Page 140: Courtesy of Éditions du Seuil, Paris. Digital image by Autumn Noble.

Page 145: Courtesy of David Brill, photograph © 1994 David L. Brill.

Page 146: (left) Photograph courtesy of David Lordkipanidze; *(right)* Photograph courtesy of F. Clark Howell. Digital image by Michael Zimmernan.

Page 150: Redrawn from figure 2 in Sewell Wright (1940). Digital image by Michael Zimmerman.

Page 154: (right, left) Maps are redrawn from figures 1 and 2 in Djubiantono and Sémah (1993) in *Pithecanthrope de Java (Les Dossiers D'Archaeologie,* no. 184). Digital images by Will Thomson, Armadillo Arts, Iowa City, Iowa.

Page 158: Adapted from figure 3 in Heslop et al. (2002) *Palaeogeography, Palaeoclimatology, Palaeoecology* volume 185. Courtesy of D. Heslop.

Page 163: Redrawn from figure 3 in Larick and Ciochon (1996) *American Scientist* volume 84. Digital image by Will Thomson, Armadillo Arts, Iowa City, Iowa.

Page 165: Redrawn from figure 2 in Rightmire (1998) *Evolutionary Anthropology* volume 6. Thumbnail photographs are digital images from casts or scanned images from field photos taken by Russell Ciochon. Digital layout by Michael Zimmerman.

Page 177: (top) Courtesy of the Institute of Vertebrate Paleontology and Paleoanthropology, Chinese Academy of Sciences; *(bottom)* Digital photograph by Michael Zimmerman of an American Museum of Natural History cast.

Color Illusrations (following page 76):

Plate 1: (top) Courtesy of David Brill, photograph © 1996 David L. Brill; *(bottom)* Courtesy of Zhoukoudian Museum, Institute of Vertebrate Paleontology and Paleoanthropology, Chinese Academy of Sciences. Digital image by Michael Zimmerman.

Plate 2: Courtesy of Russell Ciochon, photograph © 1996 Russell L. Ciochon.

Plate 3: Digital image by Nathan Totten under the supervision of Noel Boaz and Russell Ciochon.

Plate 4: (top): Illustration created by Bruce Scherting and Russell Ciochon, background image scanned and redrawn from artwork on page 81 of Savage and Long (1986) *Mammalian Evolution: An Illustrated Guide; (middle, bottom):* Illustrations by Will Thomson, Armadillo Arts, Iowa City, Iowa; illustrations © 2001 Russell L. Ciochon. Digital image by Michael Zimmerman.

Plate 5: (top) Image created by Bruce Scherting and Russell Ciochon, image © 2001 Russell L. Ciochon; *(middle, bottom)* Photographs by Russell Ciochon and Noel Boaz. Digital image created by Michael Zimmerman and modified by Nathan Totten.

Plate 6: Courtesy of Zhoukoudian Museum, Institute of Vertebrate Paleontology and Paleoanthropology, Chinese Academy of Sciences. Digital image by Michael Zimmerman.

Plate 7: Watercolor painting by Will Thomson, Armadillo Arts, Iowa City, Iowa. Image © 2001 Russell L. Ciochon.

Plate 8: Main figure is modified after figure on page 45 in Baba (1996) *Reviving* Pithecanthropus. Courtesy of M. Baba. Life-like reconstruction of *Homo erectus* (inset figure, top left) from a painting by Jay Matternes first used as the jacket illustration for *Java Man* by C. Swisher, G. Curtis and R. Lewin (2000). Courtesy of Jay H. Matternes © 2000. Layout of digital image by Michael Zimmerman.

Index